Raising
Poultry
Successfully

D0957979

Raising
Poultry
Successfully

Will Graves

WILLIAMSON PUBLISHING
CHARLOTTE, VERMONT 05445

Library of Congress Cataloging in Publication Data

Graves, Will.
  Raising poultry successfully.

  Includes index.
  1. Poultry. I. Title.
SF487.G644 1985 636.5'0068 85-6541
ISBN 0-913589-09-8 (pbk.)

Cover and interior design: Trezzo-Braren Studio
Typography: Villanti & Sons, Printers, Inc.
Printing: Capital City Press

Williamson Publishing Co.
P.O. Box 185
Charlotte, Vermont 05445
1-800-234-8791

Manufactured in the United States of America

12 13 14 15 16 17 18 19 20

Contents

Continued on next page

Contents (continued)

Part One:
Chickens

Chapter 1

WHY
RAISE
CHICKENS?

Some people raise rare and fancy chickens as a hobby, sometimes for shows. Others raise certain breeds for their hackle feathers (neck feathers), which they use to make fishing lures.

However, the best reasons that I can think of for raising chickens are to provide fresh eggs and delicious meat for you and your family.

Have you ever tasted a really fresh egg? Sometimes, when people experience a happy occasion, they exclaim, "My cup runneth over!" But, when you crack an egg into the frying pan, you do not want it to runneth over. That's an unhappy occasion. A fresh egg does not runneth over the pan. The higher the yolk stands and the more compactly the white stays together, the fresher the egg.

If you have never eaten a fresh egg, that is, an egg served on your plate on the same day it was laid, you are in for a special treat. Once you have tasted a really fresh egg, you can never go back to the supermarket variety, which may have been packaged 30 days before you brought it home.

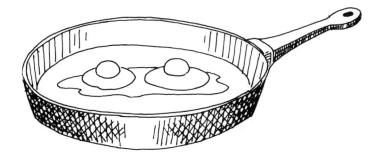

Fresh eggs look different and taste better than old eggs. A fresh egg will have a high-standing yolk and a compact white. The egg you buy in the supermarket may be a month old—notice the flat yolk and runny white.

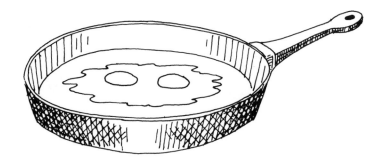

Not only can chicken be prepared in 101 different ways, but health practioners are recommending that we eat more chicken and less red meat. The meat from chickens has less fat and fewer calories, and contains more protein of higher quality, than comparable amounts of beef or other red meats.

You don't need a lot of land or space in order to raise chickens. Whether you have several acres or just a big backyard, if you have some kind of outbuilding—a shed, garage, or small barn—and if your local zoning laws don't prohibit it, and if the neighbors don't protest, you can raise chickens to have fresh eggs and tasty meat for your table for a small outlay of cash and a little effort and time.

Table 1-1

COMPARISON OF COOKED CHICKEN AND BEEF

Kind of Meat	% Protein	% Fat	% Moisture	Food Energy Calories/lb.
CHICKEN (16 weeks old, roasted and boned)				
Breast (white meat)	31.5	1.3	66	625
Leg (dark meat)	25.4	7.3	67	761
BEEF (cooked and boned)				
Round steak	27.0	13.0	59	1,056
Rump roast	21.0	32.0	46	1,714
Hamburger	22.0	30.0	47	1,650

Table 1–1. (From: Poultry Meat, Ontario Department of Agriculture, Pub. 9, 1964)

The care and feeding of chickens *is* a responsibility. Before deciding to raise chickens, consider the fact that it is a 7 day a week job. As a beginner, you will probably start out with a small flock of chickens, and their care and feeding will take only about 15 minutes of your time, twice a day, every day. This means you can't go away for the weekend and leave them to fend for themselves. Also, in the interest of economy, the killing and dressing-out of meat birds should be done by you or members of your family. If you doubt that you are capable of killing and cleaning the birds, then the raising of meat birds (broiler-fryers) is probably not for you. You could still raise laying chickens for their eggs alone.

Chickens are colorful; their behaviour is interesting and amusing. Give your children a chance to help with the daily chores. It will help them develop a sense of responsibility and will give them a hand in useful, productive work. Chickens can make reasonably good pets, although they cannot be housebroken. They will come when called, fly up onto your shoulder, and eat out of your hand. But, and this is very important, do not make a pet of any bird or animal you will ultimately slaughter and eat. Just the thought of eating a pet is repugnant to most people. Treat the birds with kindness and consideration, but don't make pets of them if they are destined to wind up in the freezer.

Speaking of freezers, you will need one with the capacity to hold about 75 pounds of meat, if you begin a broiler raising venture with 25 baby chicks, as I suggest in chapter 3.

A very good reason for raising meat birds is that it can be a short-term project. You can fill that aforementioned freezer with broiler-fryers in just 8 weeks, whereas it takes about 5 months to raise a pig or lamb to good slaughter weight, and up to 1 year to raise a beef calf. And, of course, those animals cost more to start with. You aren't liable to lose your shirt with a small flock of chickens.

However, the idea that you can save money by raising chickens is *not a good reason* to do it. All things considered, including maintenance of the chicken house, cost of the chicks, the feed, necessary equipment, electricity, and your time and labor, you cannot save money by raising your own flock. You probably can buy eggs and packaged broilers at a supermarket (albeit not as fresh and delicious) cheaper than you can raise them. This is especially true for a small flock. Larger flocks may be more economical.

Economics aside, the main reasons for raising your own chickens are *quality* and *satisfaction*. Your own eggs will be fresher, and the meat will taste better. Due to a lot of static, mostly from small flock owners, there are fewer drugs in commercial chicken feed than in previous years. This means even higher quality fresh meat — even if you don't grow your own feed.

And, there is a certain satisfaction in gathering your own eggs and barbecueing your own fryers that cannot be gained in any other way. You fed, watered, cared for, and raised these birds. Now you can enjoy the fruits of your labor. You can't buy that kind of satisfaction at the market.

Chapter 2

WHAT KIND OF CHICKENS SHOULD YOU RAISE?

Once you have decided that you want to raise chickens, and determined that you have the space, a building of some sort, and that the law allows it, the next thing to decide is whether you want eggs, meat or a combination of both. Do you want to spend 8 weeks or 17 months on the venture? In the case of eggs, do you want white ones or brown ones?

Decisions, decisions, decisions! But, they are fun decisions. What kind of chickens to raise? How long does it take? How much will it cost? What can you expect in the way of production?

TYPES OF CHICKENS

Up until about 1930, chickens were kept mostly for their eggs. Almost every farm had a small flock of medium-size, dual-purpose chickens. These chickens could produce both eggs and meat. The birds were fed on dinner table scraps, plus whatever grain the farmer's wife could scrounge from the oat bin or corn crib. Nothing very scientific about that. After a couple of years, when the hens didn't lay enough eggs to pay for their keep, they were thrown into the pot for soup or stew. When the roosters, having been fed on a catch-as-can basis, finally reached a reasonable weight, they provided the customary Sunday dinner.

Early in the 1930s, several major changes took place in the production of chickens for both eggs and meat. The practice of keeping the old-fashioned, medium-weight, dual-purpose chickens began to decrease. Poultry farmers began to specialize in raising birds for either eggs or meat.

If, up until the 1930s, cooking the meat of chickens was usually only an offshoot of keeping a flock for eggs, by the early 1930s, things had changed. Boy, how they changed! The broiler industry was booming. Heavy varieties of chickens were being used, mostly hybrid crosses of the Cornish and White Plymouth Rock breeds, to produce fast-growing birds with broad breasts, big legs and thighs, and rich yellow skin.

In 1934, about 30 million broilers were produced in the United States. In 1983, over 2 billion broilers were raised. In the 1930s, it took about 5 pounds of feed to put 1 pound of gain on a broiler chicken over a period of 4 months. In 1952, the amount of feed required to raise a broiler to about 4 pounds had been reduced to 3½ pounds, over a period of 12 weeks. Now, in the 1980s, it only takes 2 pounds of feed to produce 1 pound of weight gain on a broiler chicken.

Similar specialization has taken place in the production of light-weight egg-producing chickens, with all the brooding and mothering instincts bred out of the hens. As a small flock owner, you too can specialize in meat or eggs. Or you can raise the old-fashioned, dual-purpose chickens from which you can get both eggs and meat. There are differences in terms of time and work involved. Lets' take a look at what the differences are.

MEAT CHICKENS

Meat birds come in a few different categories. When ready for eating, a broiler-fryer is a bird less than 3 months old, male or female, with a pliable, smooth-textured skin and tender meat. The breastbone cartilage is quite flexible, in contrast to a 1-year old bird whose breastbone tip is hard and inflexible.

It used to be that broilers and fryers were prepared differently for the retail trade, thus the different terms. Broilers were sold whole or cut in half. Fryers were quartered so that there were 2 pieces combining beast and wing, and 2 pieces combining drumstick and thigh. Today, however, you will find whole birds labeled as fryers at your supermarket. It is a matter of semantics; the words broilers and fryers are used fairly interchangeably.

Although these meat birds reach good slaughter weights quickly, mature hens lay few eggs, and feeding a heavy hen for her sparse output of eggs can drive you to the poorhouse. A broiler-fryer chick can reach 4 pounds, live weight, in 8 weeks and yield about 3 pounds of edible meat, with some bone included.

A roaster is a larger chicken, of either sex, which is usually slaughtered at 5 months or less, weighing 5 to 8 pounds. They have tender meat and flexible breastbone cartilage. Capons are castrated male chickens, raised to a larger size than broiler-fryers and used for roasting.

I recommend that beginners concentrate on raising broiler-fryers for meat birds.

EGG LAYERS

The strictly laying type of chicken starts producing eggs about 5 months from the day she is born. In her first laying cycle, which can last 12 to 14 months, she can produce 20 to 22 dozen eggs. Although she can lay almost 10 times her own weight within a year, this hen is small, skinny, and nervous, and won't provide much meat for a Sunday dinner when she has outlived her productiveness.

DUAL-PURPOSE CHICKENS

Dual purpose chickens, larger than the strictly egg-laying types, are good for both eggs and meat, but they take a little longer to mature. The hens start laying at 5½ to 6 months. With some notable exceptions, they generally lay fewer eggs than the strictly egg-laying types—perhaps 18 to 20 dozen during a laying cycle. They also cost more to keep because they eat more feed; but the hens can provide a tasty dinner after their egg-laying cycle is over. With some exceptions, the cockerels (males) take longer to reach good broiler-fryer weight. However, they can be carried on to a good roaster weight of 5 to 8 pounds.

Medium-sized or dual-purpose chickens are not shown in Table 2–1. There are too many variables involved. The hens take longer to reach the egg-laying stage than egg-layers, and their egg production is usually less. The cockerels, with the exception of certain strains, take longer to reach optimum broiler weight than most meat breeds, although they can be carried on to reach good roaster weight of 5 pounds or more. Both the hens and cockerels provide tenderer meat than the lightweight, strictly egg-laying types of chickens.

FURTHER COMPARISONS OF CHICKENS

As you can see from Table 2–1, it doesn't take long to raise broilers and fryers. Raising chickens for egg production takes a lot longer and is more involved.

Although the care and feeding of any baby chick is about the same for the first 6 weeks, after that point, meat birds and potential egg layers go separate ways. The meat chick is kept on a high-protein feed ration. The egg chick is fed a ration lower in protein because too much protein at that stage of her life can cause a female to come into egg production too soon, resulting in fewer and smaller eggs and possible damage to her internal organs.

Table 2-1

COMPARING BROILER-FRYERS TO EGG LAYERS

Type of Chicken	Estimated Cost Per Chick	Time Involved	Estimated Feed Consumption	Estimated Feed Cost	Edible Meat	Egg Production
Broiler-fryer	$.57 (minimum purchase 25 chicks)	8 weeks	8 lbs./chick	$.88/chick	3 lbs./chicken (some bone included)	————
Egg-layer	$.90 (minimum purchase 15 chicks)	17 months (5 months growth plus 12 months laying cycle)	108 lbs./chick	$11.88/chick	————	20 dozen/bird (12-month laying cycle)

Table 2-1. In this table, we are comparing the costs of raising day-old Cornish-Rock chicks of mixed sexes (broiler-fryer) to day-old White Leghorn female chicks (egg-layers). Use the information to make general comparisons.

Thus, it is not advisable to keep broiler-fryers and egg layers in the same pen, certainly not past 6 weeks. Also, egg layers enjoy roosts and *must* have nests to lay their eggs in, pieces of furniture totally unnecessary for the meat birds.

With regard to time involved, most people spend more time attending to household (nonproducing) pets, such as cats and dogs, than is required in caring for a little flock of chickens.

BREEDS OF CHICKENS

There are about 200 varieties of chickens, and they come in all kinds of sizes, colors, and shapes. They vary in size from a fancy, multicolored, 1 pound bantam to a Giant Brahma which can weigh 12 pounds. Besides white chickens, there are all black, greenish black, black barred, red, brown, buff, golden, silver laced, silver penciled, and blue chickens.

Some breeds of chickens have fascinating feather patterns. The Cochins have fluffy feathers running all the way down their legs to their toes. Others, like the Turken, have naked necks. And some, like the Golden Polish, have hoods on their heads, like little top hats.

The Rock Cornish cross can gain 4 pounds of flesh in 8 weeks. With its big breast and thick thighs and legs, it is a favorite for broiler-fryer production. Mature hens are poor layers of brown eggs.

Barred Rock and Wyandotte roosters produce hackle feathers that bass fishermen use for tying fishing lures. Both of these varieties are dual-purpose chickens, good for producing both eggs and meat. The hens lay large brown eggs.

The White Leghorn is the workhorse of the egg-laying breeds of chickens, capable of laying more than 20 dozen large white eggs per year.

The Rock-Cornish, or Cornish-Rock, is favored by the large commercial broiler factories because it develops quickly. This broiler chicken has a meaty breast, thick thighs and legs, and can gain 4 pounds in 8 weeks. The mature hens are poor layers of brown eggs.

Of course, the breed of chicken you will raise depends on what you want the chickens for: meat, eggs, or both. Then you should narrow down your choices to pick a breed that appeals to you, as far as color, size, and shape.

The White Leghorn is the mainstay of the egg-laying varieties, capable of laying more than 20 dozen white eggs per year.

LEARNING MORE ABOUT CHICKENS

The problem is, how do you know what you like if your association with chickens is very limited? First, visit chicken farms and hatcheries in your area. You can ask your feed store dealer for names and addresses of local people who have small farm flocks. When you visit these people, ask them why they raise a certain bird, how much it costs, how long it takes for the birds to reach maturity, and the potential production of their flock in the way of eggs or meat. Also ask about any particular problems relative to the breed they raise.

If you can't locate any local small flock growers to visit, write to distant hatcheries and request their color catalogs. You can get the names and addresses of hatcheries from farm magazines, rural newspapers and the county extension service. Your state or county extension poultry specialist is, by the way, one of the best sources of information of all. (See the appendix for listings of extension poultry specialists, state by state, and for a partial listing of hatcheries.)

After you have visited chicken growers, talked to your poultry specialist, and scanned the gorgeous color catalogs from faraway hatcheries, you'll be more competent in choosing a variety of chickens to raise. Table 2–2 can help you narrow down the choices further. Remember, it's not much fun raising white-feathered chickens if you really like red, brown, or black birds. Raising a small flock should be fun.

WHAT ABOUT RAISING DUAL-PURPOSE CHICKENS FOR BROILERS?

Largely ignored in recent years by the big commercial broiler factories and egg producers, medium-size chickens that provided both meat and eggs were the mainstay of the poultry farmer up until the 1930s and 1940s. Examples of this type of chicken are Rhode Island Red, New Hampshire, Buff Orprington, Silver Laced Wyandotte, Barred Plymouth Rock, and Black Australorp.

Certainly one of these breeds could have been chosen to crossbreed with the Cornish Game chicken to produce a fast-growing meaty chick, just as well as the White Plymouth Rock. Except for 1 reason. Dual-purpose chickens are all birds of color with the exception of the White Plymouth Rock. At 8 weeks of age, when dressed out as a broiler-fryer, they can have tiny dark spots in the skin where their pinfeathers were removed. There is no such problem with the white-feathered White Plymouth Rock.

By the way, birds of color, once they have reached maturity and have their final plumage, will dress out very nicely, with no troublesome pinfeathers. Thus, they make excellent roasters weighing 6 to 8 pounds.

It is to the credit of our independent hatcheries that they are perpetuating these dual-purpose varieties and still supply the old favorites to the small flock grower.

HOW DO THE MEDIUM-SIZED BIRDS
STACK UP AGAINST THE HYBRIDS?

For the most part, the straightbreds will take longer to reach optimum broiler-fryer weight. So it will cost more to feed them. There are exceptions, however. Some New Hampshire strains can match the Cornish-Rock hybrids, inch for inch and pound for pound. The big broiler factories don't push standardbred chickens because hybrids tend to be more uniform with regard to growth rate and livability, and there is less chance of undesirable recessive characteristics.

Table 2-2

Heavy Breeds for Broilers and Fryers: Fast Growing

Cornish X Rock

Various crosses and hybrids of Cornish Game Chickens and White Plymouth Rocks

Heavy Breeds for Roasters: Slower Growing

Jersey Black Giants	Buff Brahmas
Jersey White Giants	White Cochins
Light Brahmas	Black Cochins
Dark Brahmas	Buff Cochins

Dual-Purpose Breeds for Eggs and Meat

Rhode Island Reds	Barred Rocks
New Hampshires	Partridge Rocks
Wyandottes	Buff Orpringtons
White Rocks	Black Australorps
Buff Rocks	

Egg-Laying Breeds: High Producers

White Leghorns	Blue Andulusians
Red Leghorns	California Grey's (hybrid)
Brown Leghorns	California White's (hybrid)
Silver Leghorns	Golden Comets (hybrid)
Black Minorcas	Production Reds (hybrid)
Buff Minorcas	Hisex White (hybrid)
Anconas	Dekalb XL Link (hybrid)

Table 2–2. Breeds of chickens, grouped by function and characteristics.

WHAT ABOUT THE VERY HEAVY BREEDS?

Jersey Black Giants, Jersey White Giants, Light Brahmas, Dark Brahmas, Buff Brahmas, White Cochins, Black Cochins, and Buff Cochins all fall into the category of heavy breed. These breeds and varieties are slower-growing than either the hybrids or dual-purpose types; but when fully developed, they have massive frames and make wonderful roasting chickens. They reach mature weights of 10, 12, or more pounds.

Why choose a slower-growing breed of chicken? They have quiet dispositions and are easy to manage. Due to their small comb, giant size, and heavy plumage, they can survive severe winter weather conditions. It would be unrealistic to raise these varieties in the hope of producing a 4-pound broiler in 8 weeks. And you wouldn't keep them for egg production as the hens are poor layers of brown eggs. They are a good choice for hobbyists, poultry fanciers, and those people who just like the way they look and act.

Barred Rock roosters are prized by fishermen for their hackle feathers, which are used to make fishing lures for bass. This is considered a dual-purpose breed, good for both its brown eggs and meat.

As a beginner, you should plan to start out small. Keep your venture into chickens within manageable bounds.

For those of you who are in a hurry to put meat on the table, chapter 3 offers a plan to raise 4-pound broilers in 8 weeks on 8 pounds of feed.

But those of you who are not in such a hurry can take things a little easier. I suggest buying about 25 day-old chickens. With that small number, it won't cost you an arm and a leg to raise them; the task of butchering them won't be overwhelming (you should do it yourself); and you can realize 75 pounds of meat from the project. It is not a long-range venture. If you don't like growing broilers, they'll be gone in a couple of months.

And I don't recommend that you go out and invest in a lot of cages, either. I can sympathize with commercial poultry producers who deal with thousands of broilers, whose living depends on programmed efficiency, and to whom time is of the essence. But, for small flock owners, not involved in commercial trade, who eat what they raise, cages and total confinement of the chickens is not necessary. I prefer the natural way: a rooster crowing at false dawn, broody hens and chickens running free—well fairly free, at least on fenced range. When you're a kid, bare feet on fresh chicken droppings is ok. When you get older, it leaves something to be desired and you build a fence.

If it is eggs you are primarily interested in, then I suggest that you start with a flock of about a dozen pullets. Keeping any more hens than this will usually result in a surplus of eggs during their peak laying season. Although you can sell, trade, barter, freeze, or use surplus eggs in baking, it often takes more effort than it is worth to dispose of the surplus.

I recommend to novices that they start with a dual-purpose breed, as they have quieter dispositions than the lightweight breeds, and they give you a roast as a bonus. Even though they eat more feed, start laying eggs later, and may provide fewer eggs, I believe the advantages of raising dual-purpose hens outweigh the disadvantages.

Chapter 3

RAISING BROILER-FRYERS FOR MEAT

Raising broiler-fryers for meat provides a quick return on your money, time, and effort. With the right variety of chickens, the right feed, and correct care, you can aim for a 4-pound bird on 8 pounds of feed in 8 weeks. You can figure that a 4-pound bird dressed will yield about 3 pounds of delicious meat (this includes bone).

WHAT'S INVOLVED?

First, you are going to have to decide whether you want to start with eggs that you incubate yourself, or whether you want to start with day-old chicks. If you are going to incubate the eggs, you will have to purchase an incubator and a candling device, or locate some broody hens and make a candling device yourself.

Then you will have to decide which breed to raise, and order either the eggs or the chicks.

Before those eggs or chicks arrive, you need to build a chicken house, either from scratch or by renovating an existing building. You will have to equip it with feeders, waters, and a heat source. Next, you need to lay in a source of feed. Then, you will finally be ready for those chicks or eggs to arrive. But, it really isn't as much work as it sounds.

STARTING YOUR BROILER-FRYER FLOCK

The best time to start your flock is from March until June. Then the weather is on your side. Day-old chicks, whether you buy them live or incubate them yourself, will need a source of heat for the first 6 to 8 weeks of their lives. Your costs will be lower if you wait for the warm weather. Although in this day and age, raising broilers commercially is a year-round business, you will need all the help you can get from Mother Nature. If you live in the Sunbelt, buy your eggs or chicks in March or April. If you are located in the North, wait until May or June to begin your flock.

Incubating Eggs Versus Buying Day-Old Chicks

How do you want to start the flock? Do you want to incubate the eggs yourself, or do you want to start with day-old chicks?

The egg incubation process takes 21 days and is full of perils, whether you use natural incubation methods (setting the eggs under a broody hen) or artificial incubation (in a relatively low-cost incubator).

First, there is the problem of obtaining hatching eggs of the breed of chicken that you want. Unless you live within a reasonable drivng distance of a hatchery that will sell fertile eggs of the variety you desire, the eggs will be shipped to you via Air Parcel Post. The shipping costs will be more expensive than for live chicks as 1 egg weighs more than 1 day-old chick.

Only a small number of hatcheries will sell eggs for hatching. Most hatcheries that *do* sell eggs for incubating only will sell them in minimum lots of 50 or 100. *All* of the hatcheries that sell hatching eggs issue a disclaimer that they will not guarantee either the fertility or hatchability of the eggs. In contrast, any reputable hatchery will guarantee 100 percent live arrival of baby chicks, or refund your money.

Hatcheries that do sell fertile eggs and will ship them, charge from $.50 to $1.00 per egg, with no guarantee of results. I have had the personal experience of having only 50 percent of the eggs I incubated hatch out. The national average for successful hatching is about 60 percent, although the makers of artificial incubators claim an average of up to 70 percent hatching success with their machines. I know of some small-flock poultry farmers who have experienced total failure in incubating eggs. Twenty-one days of anxious care and then, zilch!

If you want to put meat on your table in as short a time as possible, you will want to start with day-old chicks. If you order from a reputable source, your chicks are guaranteed to arrive alive. And you will avoid the expense of incubating machines and the fuss of watching over the incubation process.

There is, however, 1 strong reason to consider incubating eggs. Although the incubation process, whether by natural or artificial means, seems like a big hassle, I would like to say that it is *not* my intention to discourage anyone from incubating eggs. Although it is usually touted as a good project for rural children, 4H club members, Future Farmers of America, and science classes, there is no reason an adult cannot enjoy the experience of incubating eggs. The miracle of birth never ceases to amaze and delight me, whether it's the birth of a chick, duckling, gosling, turkey poult, piglet, lamb, calf, or colt.

Whether you are incubating or starting with chicks, if you are a beginner, I suggest you start with 25 broiler-type chicks. This means that to be on the safe side, you will need 4 dozen fertile eggs, assuming that about 60 percent of them will hatch out live. Four dozen fertile eggs, at the minimum price, will cost about $12.00 with no guarantees. You can buy 25 day-old broiler chicks for about $14.25, guaranteed live arrival at your local post office. Be sure the fertile eggs you buy are for meat birds. A Cornish-Rock cross is recommended.

It is my intention to provide alternatives. If you are one of those people who are in a hurry to put meat on the table, turn straight to page 33 and read about buying live chicks. Otherwise, read on about incubating eggs.

INCUBATING EGGS

If you do choose to incubate eggs yourself, there are 2 ways to go: use a broody hen to incubate the eggs naturally or buy an artificial incubator.

THE NATURAL METHOD OF INCUBATION

If you choose to use a broody hen to incubate the eggs, you have to be careful of which breed you pick. A broody hen is a moody hen. Although no one knows exactly what she is pondering or contemplating, she has to be of a mind to dwell continuously, for 21 consecutive days, on the eggs. Strictly egg-laying types like the Leghorn are too nervous and fidgety to set eggs. In keeping with our current quest for egg-laying machines, the motherly qualities have been bred out of the Leghorn and her peers. Heavy breeds like the Brahmas and Jersey Giants also are not too keen on setting eggs.

Your best choice for a broody hen is a dual-purpose hen from a medium-size variety, such as a Rhode Island Red, Orpington, Plymouth Rock, Wyandotte, or New Hampshire. These hens have calmer dispositions than Leghorns, and they are not so big that they break the eggs they set. Game chickens, so called because they originated from Indian (Asian) Games and Old English Games, often have broody dispositions; but the hens sometimes have 1½-inch spurs on their legs which can break a lot of eggs.

I have found the bantam hen to be the best mother: fierce, protective, and dedicated. Being small, she can cover fewer regular-size hen eggs, perhaps only 6 or 7. Thus you will need more than the normal 5 hens for the job. They will, if you can find them, cost at least $2 per hen.

Scout around your area for older broody hens. Choose those with proven setting tendencies, not the young, untried hens. To brood 4 dozen eggs, you will need 5 medium-size hens (or more banties), figuring that each hen can cover 9 or 10 eggs and do the job right. The hens should be free of lice and mites.

One advantage of a hen over an electric incubator is that you don't have to worry about a power failure during a bad storm. On the other hand, there is no warranty that a hen won't get up off the nest some morning, bored of sitting on eggs, and decide she'd rather chase butterflies, leaving the half-formed embryos to rot.

In order to hatch, fertile eggs must be maintained at the correct temperature, correct humidity, have adequate ventilation, and be regularly turned. Fortunately, it is the instinct of a broody hen, whether Rhode Island Red or bantam, to take care of these most important duties for the 21 days it takes an egg to hatch out. In their dedication, broody hens have been known to forget about eating or drinking or even stretching their legs, so you should provide feed and water in close proximity to the nest. She should also be protected from predators who relish eggs, such as raccoons, skunks and the family dog. Above all, don't lift her up to look underneath and see what's happening with the eggs. When the time comes, she'll let you know, and so will the chicks.

Here is the approximate cost of incubating 4 dozen eggs the natural way in order to achieve a hatching success of about 60 percent:

```
48 fertile eggs . . . . . . . . . . . . . . . . $12.00
 5 broody medium-size hens . . .   10.00
                              $22.00  Total (and no guarantee)
```

ARTIFICIAL INCUBATION

In ancient times, primitive mud-walled houses, complete with wood-burning ovens to provide heat, were used to incubate eggs. Artificial incubators were developed in the United States in the 1840s. However, it was still common practice up until about 1930 for fertile eggs to be placed under broody hens. With the advent of the artificial incubator, thousands of eggs could be set at a time, instead of depending upon hens to hatch out their small clutches.

Incubating Machines

While the eggs are incubating, you must provide constant heat and humidity for the developing embryos. You must also provide sufficient oxygen or the embryos will suffocate. Finally, you must turn the eggs 180 degrees at least 4 times a day. The purpose of turning the eggs from side to side is to prevent the embryo from adhering to the membrane on the inside of the shell, which it may do if allowed to stay in 1 position for too long. Eggs should be turned at regular hours; that is at the same times each day.

For as little as $30 you can buy a basic home incubator that will hold up to 70 chicken eggs. The electric-powered economy model comes complete with a heating element, thermostat, thermometer, and a built-in water pan to supply humidity.

This type of incubator is called a still-air machine. It depends on gravity for the flow of air through openings at the top and bottom of the incubator. These incubators are fairly successful when incubating small numbers of eggs at a time. For another $10 you can get the same model with a clear plastic window so that you can see what's going on inside the incubator at all times, plus an automatic egg-turning device, which has got to be worth the extra bucks.

A small capacity, forced-air machine costs more, with economy models starting at about $120; but the hatch ratio is much better, perhaps as high as 70 percent. The forced-air incubator has a fan for circulating air throughout the whole machine and thus provides a more consistent environment for all the eggs within.

When using an artifical incubator, always follow the manufacturer's instructions carefully.

It is important that the correct temperature be maintained at all times. With a still-air incubator, the temperature should be kept at 100 to 102 degrees F. For forced air machines, the temperature can be lower, from 99 to 99.5 degrees F.

To start, the relative humidity within the incubator should be around 85 percent. During the last 3 days before hatching, it should be increased to about 95 percent. In most small incubators, humidity is provided by water evaporating from evaporation pans. These pans should be kept full at all times. When you add additional water to the evaporation pans, add warm water to avoid lowering the temperature of the incubator.

If your machine does not automatically turn the eggs, remember that you must do it yourself, faithfully. Some people suggest that you mark each egg with an X on one side and an 0 on the other. When you turn the eggs from side to side make sure all the eggs have the same letter on top. That way you will know that you have turned every single egg. You can stop turning the eggs 3 days before the expected hatch (day 18).

Egg turning is a real hassle. I recommend buying an incubating machine that will do the turning automatically.

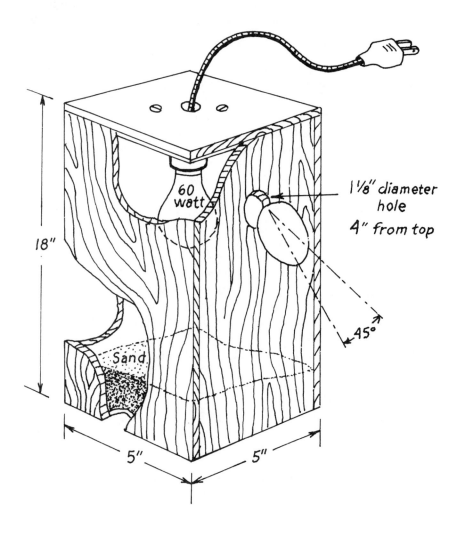

A homemade egg candling machine is easily made from plywood.

CANDLING EGGS DURING INCUBATION

Eggs are candled during incubation to see if they are fertile. The egg is placed between your eye and a source of light and examined for embryonic growth. The original source of light used in olden days was a candle. Thus the term, candling.

Egg Candling Machines

Egg candling machines start at about $25. I have always used a homemade candling device and been satisfied with the results. It is true that using a homemade device, you are likely to miss a lot of minute detail. However, you will be able to determine whether the embryo is living or dead—and that is all you really need to know. My device is quite simple. I use a 100-watt bulb and a cone-shaped piece of cardboard to focus light on the egg.

You can build a candling device with a 60-watt electric light bulb suspended in a plywood box. A 1⅛-inch diameter hole is cut into the side of the box, directly opposite the light bulb.

To get a view into the egg, you hold the egg at a slight downward angle, with the light shining behind the egg. Look into the egg, not at the light. You should be able to see into the egg to determine if the embryo is alive.

For best visibility, candle in a dark room.

What To Look For When Candling An Egg

The experts differ in their opinions as to how soon to begin candling eggs. Some say to begin on the second day of incubation, some suggest candling after 10 days and others don't candle till the 17th day, or about 4 days before the eggs are due to hatch.

What you are looking for when you candle an egg is evidence of active life within the shell.

Although within 72 hours of incubation the embryo already has a structure of blood vessels extending out from the germ spot, without the use of highly technical equipment, a beginning egg-candler won't be able to tell much during the first week of incubation. By the tenth day, however, even a novice can see if there are changes taking place within the egg.

If the egg is fertile, the egg will be dark in color, except for the air cell at the large end. At this stage, the embryo will take up about 20 percent of the space within. At 20 days, the embryo takes up almost all of the space except for the air cell.

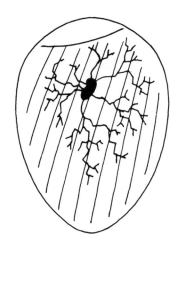
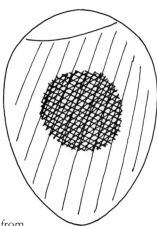

Candling enables you to separate the fertile eggs from the infertile ones. The egg on the left is fertile, revealing a small embryo and a network of veins. The egg on the right is infertile and should be removed from the incubator.

Within the first 24 hours of incubation, the eye begins to form in the embryo. Within 2 days , the heart begins to beat. Legs and wings begin to form on the third day. By the sixth day, the beak and egg tooth begin to form. On the thirteenth day, claws appear. On the fourteenth day of incubation, the embryo's head turns towards the large end of the egg (where the air cell is). On the seventeenth day, the beak turns to the air cell. The embryo begins to ingest the yolk on the nineteenth day and completes the job on the twenty-first day.

I think it is sufficient to candle twice during the artificial incubation period, at intervals of a week apart. Begin on the tenth day, and repeat again on or about the seventeenth day.

Do not candle after the seventeenth day. At this point, the embryo is preparing to ingest the remaining yolk, become a chick, and hatch out. This is a critical time and exposing the egg to drastic temperature changes or rough handling can disturb the whole process and destroy the embryo. If the incubator you are using does not have

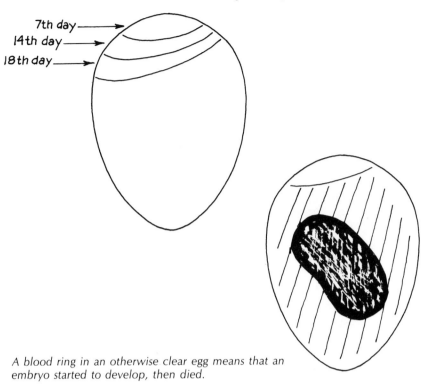

The air cell grows larger as incubation progresses.

7th day ⟶

14th day ⟶

18th day ⟶

A blood ring in an otherwise clear egg means that an embryo started to develop, then died.

a window for observation, stifle your curiosity, and don't keep opening the door to peek in and see what is happening. What can happen is that you may well lose some of the hatch due to the sudden introduction of cold air.

If the egg is infertile, it will appear fairly clear inside, except for a grayish mass (the yolk), which will move easily from side to side when you turn the egg. Infertile eggs should be removed from the incubator and disposed of. A spoiled egg can serve as a source of bacteria and contaminate the other healthy eggs.

I have never candled eggs being set by broody hens, not wanting to upset them and preferring to let nature take its course. Some hens will kick infertile eggs out of the nest. Others will sit optimistically, full term, on a whole clutch of dead eggs.

After the chicks have hatched, they should be left in the incubator for 6 to 12 hours, until they are all dried off and moving lively. At this point, they should be removed from the incubator and placed in brooding facilities. More about brooding later on.

WHAT WENT WRONG?

If your candling operation reveals a lot of infertile eggs, or if your chicks are not healthy and viable, try to find the probable cause and remedy the problem before attempting another batch of eggs. In most cases, the problem is improper temperature or lack of ventilation—operator errors that are easily adjusted.

Problem	Possible Cause
Eggs clear—no blood ring or embryo growth	Improper mating; eggs too old; hens not in good condition
Eggs clear—some evidence of embryo	Incubator temperature too high; eggs chilled; hens not in good condition or not fed proper rations
Embryos dying at 12–18 days	Lack of ventilation; hens not fed proper rations
Chicks dead without pipping	Wrong temperature; poor heredity; improper turning of eggs
Chicks dead in shell	Improper temperature or humidity
Shells stick to chicks	Humidity too low during incubation or hatching
Chicks smeared with eggs	Temperature too low; humidity too high
Small chicks	Eggs too small; humidity too low; temperature too high
Chicks' down too short	Temperature too high; humidity too low
Hatching too early	Temperature too high
Hatching delayed	Temperature too low

BUYING BROILER-FRYER CHICKS

If you decide you'd prefer to start out with day-old chicks rather than incubate eggs, buy the finest quality chicks that you can find of the right breed or variety. You are looking for a type of broiler chick that has been scientifically bred for its fast gain and tender meat with good finish characteristics.

The best chick for this purpose is a Cornish-Rock cross. This chick is the result of mating a Cornish male with a White Plymouth Rock female. These crosses have lots of hybrid vigor, or heterosis, which means an increased capacity for growth.

Some of the varieties offered to you will have names such as Hubbard, H & N, Shaver, or Vantress broilers. These are tradenames of the prime breeders of foundation stock.

It should be mentioned that you can carry these chicks past their broiler-fryer weight. When they are 10 or 12 weeks old, these chicks can weigh 5 to 6 pounds, which is good roaster weight. Ultimately, a Cornish-Rock will reach a weight of about 10 pounds.

The majority of the Cornish-Rock broiler chicks offered by hatcheries will be white-feathered and have a yellow skin. There is a reason for this. Chickens of broiler age often have very small unfinished feathers, called pinfeathers, which must be plucked out in the dressing process. The pinfeathers of chickens with colored feathers can leave dark and unsightly blotches in the skin. White-feathered chickens show less discoloration when their pinfeathers are picked.

The supermarkets, anxious to satisfy and appeal to the eye of the consumer, want a bird with an attractive carcass in their display cases. The consumers indicate their preferences to the retailer, who dictates to the wholesaler, who in turn tells the prime breeder. So, the most popular commercial broiler is a white-feathered bird.

A Bargain To Avoid

Every spring around Easter time, some feed companies and farm stores offer free chicks if you buy a bag of expensive feed from them. *These baby chicks will not do the job for you!* They will not gain 4 pounds in 8 weeks on 8 pounds of feed. They are usually cockerels culled from lightweight egg-laying chickens. It can take them up to 4 months to gain 4 pounds.

Culled cockerels are much cheaper than the heavy meat-type broiler-fryer chicks. If you purposely ordered this type of cockerel from a hatchery, 25 of them would cost about $5.50 ($.22 each). Twenty-five Cornish-Rock cockerels bred for meat from the same hatchery would cost about $16.50 ($.66 apiece), or three times as much. No one gives anything away.

What To Buy

I would suggest that you buy 25 "straight-run" Cornish-Rock day-old chicks. Straight-run means "as hatched." The baby chicks are not separated as to sex; half of the chicks will be females and half males. Although males utilize protein better, gain faster, and grow larger than females, your ultimate feed costs will not be affected that much with only 25 chicks. Ordering cockerels only will cost about $3 more than ordering straight-run.

I specify ordering 25 chicks because if you order only 24 chicks they will cost more. Buying just 1 more chick can make a difference. As an example, 24 Cornish-Rock chicks cost $16.08 from a certain hatchery. Twenty-five chicks cost about $14.25. You save $1.83 by ordering just 1 more chick. This price differential is due to the fact that hatcheries normally sell baby chicks in lots of 25, 50, 100, etc. Their mode of operation is based on these numbers. Some hatcheries even charge a $3.00 fee if you order less than 100 chicks.

I also suggest that your day-old chicks be vaccinated for Marek's disease. Most hatcheries do this as a matter of routine. Marek's disease is a form of leucosis, in which the liver becomes enlarged and the bird can become paralyzed. For an extra fee, the chicks can be debeaked and dubbed (removing the comb) at the hatchery. This is to prevent cannibalism.

In debeaking chickens, about half of the upper beak and just the tip of the lower beak is removed. This helps prevent cannibalism. They can be debeaked at any age.

Personally, I don't like the way chickens look when they have been debeaked or dubbed. Terribly unnatural. If you are raising only 25 broiler chicks and they have enough space, indoors or out, I don't think it's necessary (Cannibalism can be prevented, see page 55.) If cannibalism does become a problem in your flock, you can always debeak or dub as it becomes necessary.

Meat-type chickens usually don't have combs as large as those of the egg-laying varieties. Although another bird's comb gives a macho bird something to peck, at 8 weeks of age, when your broiler chicks should be ready for slaughter, their combs will not be a big target. Combs will freeze in extremely cold weather, and this can cause a set back in a bird's growth. But, if you begin your broiler-raising venture in the spring as I recommend, this should not cause a problem.

One major problem with debeaking young chicks is that it can cause "starve-outs." A day-old chick may not be able to eat feed or drink water if part of its beak is missing. Thus it dies of starvation.

Where To Buy Your Chicks

Every state in the union and almost every county within each state has a poultry extension service specialist associated with a university. (See the appendix for listings.) The poultry specialist can direct you to a reliable source for chicks.

Ask poultry farmers in your area where they bought their birds. Rural newspapers and farm magazines carry hatchery advertisements. If there is a hatchery within reasonable distance of your home, it is wise to buy your chicks close by. The less stress and strain on the baby chicks, the better their chance to survive. However, I have had excellent results with chicks that were shipped to me via Air Parcel Post from hatcheries at a distance of a thousand miles or more.

It is important that you buy your chicks from a hatchery whose products are classified as "US Pullorum-Typhoid clean." How does a hatchery gain this status? By participating in the National Poultry Improvement Plan. This plan, under the auspices of the USDA, involves state authorities in the administration of regulations for the improvement of poultry, poultry products, and hatcheries. The N.P.I.P. provides a federal-state program that establishes standards for the evaluation of poultry breeding stock and hatchery products with respect to freedom from hatchery-disseminated diseases.

What all this means is that not only were the chickens at a particular hatchery evaluated for general production efficiency, but the parent stock of the chicks you buy were blood tested for specific poultry diseases. The USDA will send you a copy of the N.P.I.P. directory, which includes current participating hatcheries and production reports. (See address in appendix.)

Whether you incubate eggs and hatch out the chicks yourself, or buy day-old chicks, you need to have a place to put them. It doesn't have to be large or fancy. You can use any standing building on your place: and old chicken house, milk house, garage, tool shed, or barn. Or you can build a small chicken house (see appendix for plans).

If possible, the chicken house should open onto a well-fenced chicken yard. The yard does not have to be large. With a laying flock, an outside range cannot really be too large. But, you are raising these birds for meat, and while racing up and down is good for horses, it doesn't make for tender legs and thighs in broiler-fryers. A small yard 10 feet by 15 feet is better than no yard at all. A yard 25 feet square is fine. There are no hard and fast rules here. It all depends on how much space you have available, the topography of the land, and the type of vegetation that will be enclosed. The yard should be well-drained as pools of stagnant water are a potential source of disease for the flock.

The fencing should be woven wire at least 5 feet high and fastened to posts that are 10 to 12 feet apart. The posts should be set

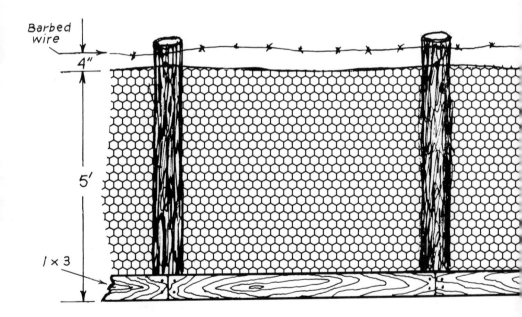

This fencing arrangement makes a fairly secure chicken yard. The woven wire fence should be at least 5 feet high. The board at the ground level helps keep chickens in and small animals out. The barbed wire at the top, a strand of barbed wire about 2 inches above the ground outside the fence (not shown), also discourages predators.

2 feet deep in the ground. A more secure arrangement would be to nail 1 × 3 boards to the posts, at ground level, on the inside of the fence. Then, run a strand of barbed wire about 2 inches above the ground on the outside of the fence, and at least 1 strand of barbed wire about 4 inches above the woven wire. This will discourage large predators.

Housing Requirements

Whatever the building, there are some very important requirements. The building has to be dry, draft-free, and varmint-proof. It should be located in a well-drained area, preferably with access to a yard. If the building has windows, it is preferable that they be on the south side. The southern exposure on chilly spring days will help keep the building warm, if you live in a northern zone. If you live in the Sunbelt, windows on at least 2 sides of the building will help to provide cross ventilation for particularly hot and humid days, once the chicks have feathered out but have not reached their optimum slaughter weight. In southern zones, an insulated roof helps to keep the building cooler. Chicks feather-out best when they have a cool area to move about in.

Ventilation. If the chickens are to be kept in total confinement, that is, in a small building with no windows, then by the time they are about 6 weeks old, you will have to provide some form of ventilation. This can be done with a combination of roof vents and fans. Lack of ventilation can cause serious disease problems. Besides, it isn't much fun to walk inside a chicken pen and have strong ammonia fumes knock your socks off. Caring for the chickens should be fun.

Temperature. Once they have feathers, chickens seem to do best when the temperature is between 55 and 80 degrees F. Fully grown chickens can take cold temperatures, but not drafts.

Protection from Predators. You are going to have to keep the day-old chicks warm, dry, and safe from predators, such as rats, raccoons, weasels, and family pets. All doors and windows must shut tightly. It is a good idea to nail a fine meshed wire on the inside of any windows so that wild birds, such as sparrows, cannot fly in and bring diseases with them. This is especially important for chicks that will be raised on litter on the floor, as I suggest you raise yours.

Space. The building does not have to be large for your 25, day-old, chicks. However, by the time the birds approach 8 weeks of age, they will need 1 square foot of space, each. You will also need floor space for feed troughs and water fountains and some elbow room to allow you to do your chores. An area with dimensions of 10 feet by 10 feet is ideal. A coop with an area of 12 feet by 12 feet, or even 10 feet by 15 feet, will work just fine.

In the beginning, you will need to erect a circular guard to confine the chicks to a small space around the sources of heat, light, water, and feed. Later on when the guard is removed, any extra space is a plus. The more room they have, the less inclined the chickens will be to pick on each other.

If you want to use a barn, but it's too large, you can partition a section of it, preferably on the south side. Again allow at least 25 square feet, and 100 square feet is ideal. If you make a coop within a barn, make sure the partition goes all the way from floor to ceiling. This will prevent drafts and keep predators out. (Rats and weasels can slip through holes of 1 inch diameter.)

Materials. The floor of the chicken house can be made of wood or concrete. Concrete is easier to keep clean, but both are adequate. An earthen floor will not do. It will be too damp, which can cause diseases, particularly respiratory diseases.

With the cost of lumber and building materials today, the construction of a new chicken house is prohibitive, unless you are sure that you will be raising broilers for a long time.

Make sure whatever structure you use for a pen is as clean as can be. First thoroughly clean the house and *then* disinfect it. You must sweep the ceiling, brush and scrub the walls, and scrape the floor down. There are a lot of good sanitizing solutions available at farm and feed stores. It is important to follow directions carefully when using disinfectants so that the baby chicks won't be harmed by any leftover chemical residues. After cleaning and disinfecting, it is best to allow the pen to stand vacant for a month, or at least 2 weeks, to eliminate all traces of the disinfecting chemicals.

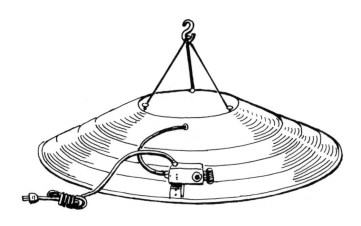

The metal hover of this electric brooder is only a few inches off the floor. If the chicks get too hot, they can escape from under the canopy.

NECESSARY EQUIPMENT

You are going to have to supply a source of heat, a waterer, a feed trough, a chick guard or corral, and some form of litter for the new chicks during their brooding period.

Heat

Baby chicks have been kept warm in many ingenious ways, including the use of hot water bottles, heated stones, and even by placing the box containing the chicks on top of dung heaps in order to take advantage of the rising heat. Along more conventional lines there are natural gas and propane gas brooders. Some are oil-fired, and there are even table-top models, in the form of metal box-like arrangements that use electricity for heat. The gas-fired and oil-fired brooders have a large hover, or cap, above the heat source, in order to confine the heat to floor level and give the chicks a place to come into from out of the cold.

You don't have to invest in anything fancy. I recommend using a 250-watt heat lamp suspended from the ceiling as the heat source. Ordinary electric light bulbs do not have the heating qualities of a heat lamp. Infrared heat lamps cost more but are preferred over the white bulb kind, as the red light tends to discourage cannibalism.

The hover brooder is heated by propane.

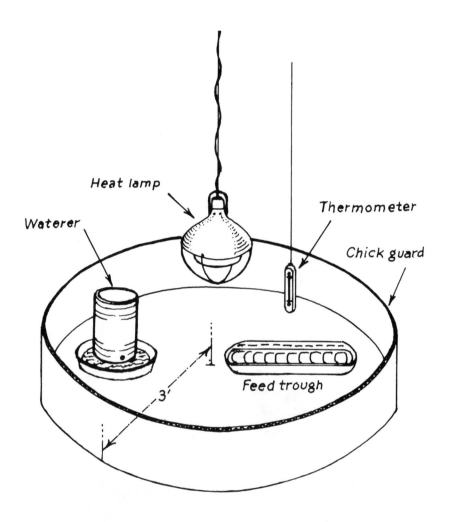

Here's a typical brooding set up. The chick guard made of corrugated cardboard keeps the chicks close to the source of heat, food, and water. Heat is provided in the form of a 250-watt infrared heat lamp suspended from the ceiling. A thermometer allows for careful monitoring of the temperature.

The lamp should be complete with a porcelain socket, a reflector on top, and a metal guard around the bulb to keep it from breaking in case it falls. Hang the lamp in the center of the pen in such a way that you can adjust the height up or down.

If the building you are using has no power, you can use a heavy-duty extension cord to bring in electricity from the nearest sources.

Waterers

When the chicks are small, a single 1-gallon waterer, either plastic or metal, will provide enough water. By the time your 25 chicks are 5 weeks of age, they will be drinking 1 gallon of water every day. At that point, you will want to add a second waterer.

The waterers need not be fancy. You can make one out of a 1-gallon can and a pan. Clean the can thoroughly. Drill 2 holes into the sides of the can, each about 3/4 inch from the lip of the can. Fill the can with water. Place the pan on top. Invert the can and pan. The water will slowly flow out of the can from the holes to maintain a constant water level in the pan.

Although older birds can be watered in open pans, pails, or troughs, young chicks must be protected from such invitations to drowning.

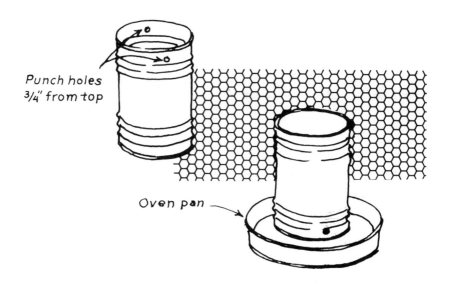

Punch holes 3/4" from top

Oven pan

This homemade waterer is made from a 1-gallon or restaurant size (No. 10) can and a plain tin oven pan. The can is filled with water, the pan is placed on top, then the whole setup is inverted.

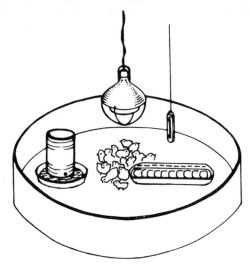

Chicks too cold. Lower lamp.

Chicks too hot. Raise lamp.

Cold draft. Plug it up.

All is well; conditions just right.

Feeders

Chicken feeders come in many styles, shapes, and sizes. What to look for is a feeder that is designed to prevent feed waste. This means the feeder should have a lip to prevent the chicks from beaking out the feed. The feeder should be adjustable in height so that as the chicks grow, the height of the feeder can be adjusted.

At first, each chick will need about 1 inch of space at the feed trough. You can buy baby chick feeders with grills or slotted openings so that each bird has a slot and none are starved out. Baby chick feeders usually are not adjustable for height, but lie flat on the ground.

As a rule of thumb, whatever type of trough chickens feed from, the lip or edge of the feeder should be about the same height as the bird's back. This enables them to reach the food comfortably without encouraging them to pull the feed out. In time, as the chicks grow, the feeder should be raised. As the chicks get larger, they will have trouble getting their heads into small slotted openings. When they are about 3 or 4 weeks old, it is a good idea to provide an open trough feeder with adjustable legs. Metal and plastic types can be purchased at farm supply stores or you can make one yourself out of wood.

A trough that is 4 feet long, 4½ inches wide, and 2½ inches deep is sufficient for 25 broiler chicks up to 8 weeks of age. The feed trough should have a roller bar on its top so that the birds cannot climb into the feed and soil it, thus causing waste.

Chick Guard

A chick guard is a corral to keep chicks close to the source of heat, water, and food. It also prevents drafts from reaching the chicks. A guard with a diameter of 6 feet is a reasonable size to start with. It will have to be enlarged as the chicks grow. The guard should be 12 inches high.

You can buy ready-made chick guards, or make one yourself out of sturdy corrugated cardboard. Set the guard in a circle around the heat lamp, feeder, and waterer. The guard should be at least 3 feet from the heat source. The guard is set up in a circle as chickens have a tendency to run into corners and climb all over each other when frightened. This can cause injury or suffocation. As the chicks grow, the circumference of the guard can be enlarged. After a couple of weeks, it can be removed from the pen.

Litter

Litter is any kind of bedding material that will absorb droppings, help reduce moisture in the pen, and serve as insulation in cool weather. Litter should be clean and dry, not moldy or dusty.

Dry sawdust, wood shavings, finely chopped straw, ground corncobs, and peat moss all make good litter. It really depends on what is most available in your area and the cost. Some farmers put down a base of dry sand underneath the litter.

The bedding should be 3 to 4 inches thick at the start. In cold weather, a thick base of litter helps keep the chicks warm. In very hot weather, a thin base works better. When the litter becomes wet or soiled by droppings, it should be removed (you can use a small shovel or dustpan), and more dry litter added. Stirring up the litter occasionally helps to keep it from packing down.

Particular attention should be paid to the condition of litter around the waterer. Damp litter can cause serious disease problems among chickens raised on the floor. Small portions of hydrated lime can be added to very wet spots in the litter in order to keep it dry. *Never use old litter from another chicken house when brooding chicks* — there is too much chance of exposing them to disease.

FEEDING THE CHICKENS

These little chicks that you are going to raise into tasty meat within the next couple of months need a starter ration of 22 to 24 percent protein in order to build strong bones, especially legs, to support the extra heavy bodies, which will gain 4 pounds in 8 weeks. Nature probably never intended for any bird to grow that heavy that fast. But, if you supply your broiler-fryer chicks the correct feed, their legs will be strong enough to hold them up.

CAGES AREN'T NECESSARY

Cages are commonly used by commercial poultry raisers, particularly commercial egg producers. Floor space requirements per bird are less with cages and cage systems can be adapted to automatic feeders and waterers, so labor is reduced. Baby chicks raised in cages are less prone to parasites and certain diseases, such as coccidiosis. However, they are subject to foot and leg problems, and their growth rate is not as good as that of birds raised on the floor.

I prefer to see a small flock of chicks raised on litter on the floor. I like to see the chicks have a little freedom, rather than be cooped up in cages. Medication in the chickens' feed can prevent coccidiosis. Many other diseases and parasite problems can be avoided by using clean litter and seeing that the litter remains dry.

Growing Your Own Grain

How about growing your own feed? Broiler-fryer chicks need a well-balanced ration of protein, calories, vitamins, and minerals. For the first-time chicken raiser, I recommend that you purchase ready-mixed starter ration for the birds. Otherwise, unless you already operate a working farm and have grain on hand, you will have to plant crops—corn, wheat, and oats—the year before beginning your broiler-raising venture.

It isn't that you need a large amount of land; actually, 1 acre or less of good, productive land could produce enough grain to feed a small flock of chickens. But, buying the necessary equipment to grow the grain is very expensive. At the very least, you need the following items of machinery: tractor, plow, harrow or disc, seeder, corn planter, cultivator, wagon, and combine or grain harvester for the oats and wheat.

Even if equipment is purchased used, these items easily could cost about $5,000. A rototiller does a marvelous job of preparing the seed bed and cultivating, but I haven't met one yet that can plow a decent furrow in thick sod. You have to plow before you can harrow and seed. Instead of a tractor-drawn seeder, you could use a hand-cranked cyclone seeder, a device which you hold in one hand and crank with the other as you walk through the field. This works fine with wheat or oats, as the seeds can be broadcast on top of the ground, but corn has to be planted below the surface. I won't even mention buying a corn picker here, as on very small acreage, you could pick the ears by hand and throw them into a wagon or pickup truck.

An alternative is to hire a custom operator to do the plowing, harrowing, seeding, and harvesting for you. If you can find someone willing to work on small acreage, it will cost from $25 to $40 per hour.

It is not my style to discourage anyone from growing their own grain. However, I have to be practical and remind you that for less than $25, you can buy enough feed to bring your 25 broiler-fryer chicks up to a good slaughter weight. And the feed ration will be mixed correctly in exactly the right proportions.

What Chick Feed Contains

A complete starter ration for the chicks should be ground into mash and contain the following ingredients in correctly proportioned amounts:

- **Yellow corn**
- **Soybean oil meal**
- **Fish meal**
- **Alfalfa meal**
- **Dried whey (the watery part of milk)**
- **Bone meal**
- **Iodized salt**
- **Vitamins: A, D3, K, B12, E Niacin, Riboflavin, Thiamine, Pantothenate, Choline, Biotin, Folic Acid, and Pyridoxine**
- **Minerals: Calcium, Phosphorus, Magnesium, Potassium, Cobalt, Copper, Iodine, Iron, Manganese, Molybdenum Selenium, and Zinc**

You couldn't mix this ration up in your own backyard; but even if you could, it wouldn't be economical to do so. Homegrown grains certainly can be used for the bulk of the ration, although you won't know the protein content of your grains unless they are tested. Wheat, barley, or oats can be substituted for the soybean meal, provided their protein percentage is high enough.

Homemade Chick Starter Ration

The following is a formula for mixing up 100 pounds of chick starter ration. The formula dates back almost three-quarters of a century.

- **30 pounds Cornmeal**
- **20 pounds Wheat bran**
- **20 pounds Wheat middlings**
- **10 pounds Ground oats**
- **10 pounds Ground bone meal**
- **10 pounds Beef scraps**

Wheat bran is the coarse ground kernels that are a by-product of sifting flour. Wheat middlings are medium-sized particles, separated in the process of sifting ground grain for flour.

If you *could* put this mix together, your chicks would probably survive, provided they didn't get coccidiosis or Marek's disease. I suspect the mortality rate of baby chicks was quite high and the growth rate rather low in those days.

If you grow your own grain, you will have to take it to a mill and have it finely ground. Baby chicks have to be fed mash to start out with, as their digestive systems aren't developed enough to handle whole kernels of grain. By the time they are about 6 weeks old, they can be fed cracked corn or other grains.

Commercial chick starter feed usually contains antibiotics, a coccidiostat, and antioxidants. The antibiotics are added to promote growth and prevent disease. The coccidiostat is a drug that prevents coccidiosis in chickens. The disease is spread by droppings. The mortality rate can be high. Chickens raised on the floor, or, so to speak, on litter, will definitely be susceptible to coccidiosis.

Although I appreciate how the antibiotics can spur the growth rate in chickens, I am not completely sold on the shotgun approach to preventative medication, particularly with small farm flocks. I do think the coccidiostat is a good idea. Indeed, the best reason to grow your own feed is to avoid the antibiotics.

Antioxidants are added to commercial feeds to prevent them from spoiling, oxidizing, and becoming rancid or moldy.

Buying Feed

Unless you do grow and mix your own feed, the next step in the process of setting up your chicken coop is to buy a 100-pound sack of high-protein chick starter ration and have it on hand before the chicks arrive.

Always store feed in tight containers. A clean garbage can with a secure lid works very well. Raccoons, squirrels, and mice love chicken feed. One rat can eat and/or spoil 100 pounds of feed in a year.

It is customary to feed broiler-fryer chicks a high-protein starter ration for the first 5 or 6 weeks of their lives and then switch to a grower or combined grower-finishing ration until they reach the desired weight. Growing-finishing rations contain less protein than starter rations, so they are generally less costly. They do not usually contain drugs that affect the quality of poultry meat for human consumption.

In any case, always follow the feed manufacturer's directions and ask your feed dealer just exactly what the feed contains in the way of drugs. This way you will know if a particular ration will have to be withdrawn for any period of time before the birds are slaughtered for eating.

PREPARING FOR THE ARRIVAL
OF THE CHICKS

Well in advance of the chicks' arrival, the chicken house should be disinfected and aired to eliminate any traces of the disinfectant. Just before the chicks arrive, or before they are due to hatch out, the heating arrangement should be tested and the brooder heated.

Setting the Heating Levels for the Chicks

Start out by hanging the heat lamp at a distance of 20 to 24 inches above the floor of the pen. During the first week of brooding, you must maintain a temperature of about 95 degrees F. at floor level.

You will need a sturdy thermometer to help you sustain the correct temperature. Position the thermometer at floor level near the lamp, but not directly under it. You can hang the thermometer from the ceiling on a wire or set it on the floor with a wire mesh guard around it, so that the chicks won't peck it apart or knock it over.

The day before the chicks are due to arrive or hatch out, turn the lamp on. Let the lamp heat the area for several hours; then check the thermometer. If the temperature is less than 95 degrees F., lower the heat lamp an inch or so. If it is over 95 degrees F., raise it a notch. You will have to go through this process several times in order to maintain the correct temperature. After each adjustment of the heat lamp, wait a couple of hours before you check the temperature again.

Some farmers use 2 heat lamps as a safety measure in case one burns out. Follow the same procedure for testing the temperature levels, using both lamps together.

Now that the chicken house is ready, the necessary equipment is set up, the heating arrangement has been tested, and you have feed on hand, you are ready for the big event. It's time to bring the chicks home from the hatchery or post office, or move them from the incubator to the brooder.

If You Pick Your Chicks Up at the Post Office

If your chicks are coming from a distant hatchery via Air Parcel Post, your postmaster will call and tell you that your baby chicks are peeping all over the mailroom. Open the crate in front of a clerk and count the live birds.

Most hatcheries include a couple of extra chicks in each shipment, in case any birds die enroute. If you are short of the full number ordered, file a claim with your post office. A reputable hatchery will replace the missing or dead chicks or refund your money. Some hatcheries will even replace at half price any chicks that die between the first and tenth day after they have arrived at your place without any restrictions.

BROODING THE CHICKS

When brooding baby chicks, you start out with a pen temperature of about 95 degrees F. for the first week, and decrease it by about 5 degrees a week. The temperature will be regulated by raising the heat lamp an inch or so each week. It will require a constant checking of the thermometer until you can maintain the correct temperature.

If you have incubated eggs and hatched out your own chicks, make sure they are dried out, fluffy looking, and vigorous before moving them from the incubator to the brooder. This may require leaving them in the incubator for several hours after they have hatched.

Just before it hatches, a baby chick ingests the remaining egg yolk within the shell. The yolk can nourish the chick for up to 72 hours after it hatches out, which is why commercial hatcheries can ship day-old chicks for long distances.

Be that as it may, the sooner they drink and eat, the better the chicks' chances for survival. When you place the chicks in their pen, dip their beaks into the clean, fresh, lukewarm water in their fountain and teach them to drink, one by one.

Even though you have a well-filled feed trough in front of them, it is a good idea to have small portions of the starter mash spread about on little flat box tops placed around the pen. In the beginning, the chicks may be awed by the feed trough. Again, introduce them to food by dipping their beaks into the feed in the box tops and trough.

Chicks are quick learners. After a few days, they will be eating out of the regular trough and you can remove the box tops from the pen.

As a general rule, feed troughs should not be filled more than halfway. However, for the first few days of feeding, I fill the trough all the way, until the chicks become accustomed to eating from it. I would rather have them waste feed by hooking it out of the trough with their bills (I don't debeak chicks), than starve to death. Better to lose feed than chicks.

Once the chicks have gotten settled into their new home, it is just as important to observe their behavior as it is to read the thermometer. If they are all clustered right under the heat lamp and chirping plaintively, it is too cold and you should lower the lamp. If they are all crowding the walls of the chick guard and panting with wide open beaks, it is too hot. Raise the lamp. If they are huddled along one side of the guard, they are suffering an unwelcome draft. Check the walls, doors, and windows of the house and plug any cracks or openings. If the chicks are spread evenly around their pen, drinking water, eating feed, peeping contentedly, and running around from here to there with the ebullience of youth, all is well and you have got it just right.

BROODING MANAGEMENT

This baby chick, this busy little ball of fluff that you are observing, is entirely dependent upon you for its survival. It weighs little more than an ounce, and its body temperature is about 107 degrees F. It breathes faster, its blood circulates more rapidly, and its digestive processes work more quickly than that of any other farm animal. It can't sweat, suffer drafts, or fend for itself.

You have to keep this chick warm and dry, feed the correct feed, supply plenty of clean water, and keep predators away. If you can manage to do this, the broiler-fryer chick will weigh better than a quarter of a pound at 7 days of age. At 4 weeks, it will weigh at least 1½ pounds. And at 8 weeks, it should weigh 4 pounds and provide about 3 pounds net of barbecued delight, or go into the freezer.

The key to good management is in setting up a regular schedule for feeding, watering, caring for , and observing the chicks—and then stick to the routine religiously. With a small flock of chickens, it will take 15 minutes, every morning and again at night, to satisfactorily take care of the chores.

You will begin by feeding the birds a starter ration of 22 to 24 percent protein. This feed is also very high in energy, containing over 1400 calories per pound.

Setting Up a Routine

Every morning when you go to the chicken house, remove any leftover feed from the trough, particularly if its appearance is questionable. Keep fresh feed in front of the chicks at all times, but only fill the trough halfway. Rinse the water fountain and fill with fresh water. Pick up all the wet or soiled bedding with a little shovel or dustpan and add fresh litter. Remove any dead chicks and incinerate or put them in a disposal pit. Repeat this process each night.

Take a good look at your flock. As hardhearted as it sounds, it is usually best to dispose of any obviously sick birds. Any chick that acts paralyzed or can't get up, whose head is drawn back into its shoulders, or who suffers from acute diarrhea or bloody droppings, should be removed from the pen and put out of its misery. Other signs of disease to watch for are spraddled legs; listlessness; and rough, stained, or frizzled feathers.

Chickens are prone to many diseases, but the one most likely to affect your flock is coccidiosis, which is why I recommend using a feed that contains a coccidiostat. The symptoms are listlessness, lack of appetite, and diarrhea. Again, remove any diseased bird from the flock at once, and dispose of it. If more than 1 bird shows symptoms of disease, consult your local extension poultry specialist. It may be advisable for you to send samples to a state laboratory for possible diagnosis and treatment.

Resign yourself to the fact that you may lose some of the chicks. The mortality rate of chickens is about 2 percent for the first 3 weeks of their lives. From then on, the mortality rate averages about 1 percent a month. These figures are based on the actuarial records, compiled over many years by commercial poultry growers who raised hundreds of thousands of birds. With a very small flock, you may get lucky and not lose any chicks.

As the Chicks Grow

When brooding chicks, it is customary to reduce the temperature in their pen by 5 degrees per week, provided the chicks are comfortable with the change.

In the second week of brooding, raise the heat lamp a notch and try to maintain a floor temperature of 90 degrees F. Again, observation is the best test; if the chicks complain, go back to where you were.

In the first 2 weeks of their lives, the chicks will double their weight and also double their feed intake. From then on, they usually remain on a steady rate of eating and gaining until about the fourth or fifth week, when there is a tremendous increase in both their appetite and body weight. I have no explanation for this phenomena.

Each week you will decrease the temperature in the pen by 5 degrees, always, of course, depending upon the comfort of the chicks. Expand the diameter of the chick guard as the chicks grow and add more litter. Eventually you can discard the chick guard. Keep the floor covered with a good bed of litter.

By the time your 25 chicks are about 5 weeks old, they will be drinking 1 gallon of water each day. As I mentioned earlier, you can either add a second water fountain to their pen, or make more frequent trips to fill the original waterer, making sure that they always have fresh, clean water available.

At this time, everything being relative, having reduced the temperature 5 degrees per week, you will be maintaining a brooder temperature of 75 degrees F., and the chicks will be feathering out. If the temperature of their pen matches the temperature outdoors, you can cut off the heat lamp. If nights are cool, put it on at night.

When the 25 chicks are 6 weeks old, they will have eaten about 80 pounds of their starter ration. It's time to buy a bag of grower-finisher ration. This feed will have a protein content of 18 to 20 percent and is less expensive than starter feed. Use whatever is left of the starter ration and mix it in with the grower-finisher ration. When feeding any bird or animal, it is not advisable to make abrupt changes that may cause severe digestive problems. Always try to make the transition from one feed to another as gradual as possible.

Remember to keep feed in front of the chickens at all times, but don't fill the trough more than half full. If possible, raise the feeder so that its edge is on a level with the birds backs.

If you have discontinued using the heat lamp and are keeping your flock in a windowless house, it it a good idea to use a low wattage bulb for interior light. One 40-watt incandescent bulb should provide enought light for you to do the chores and observe the chicks. They will still find the feed trough and waterer in the dim light. Bright lights encourage cannibalism in totally confined birds.

AVOIDING CANNIBALISM

Chickens are one of the favorite examples psychiatrists use to point out the existence of a pecking order or social ladder in society, whether animal or human. Even at the tender age of 2 or 3 weeks, some aggressive chicks will try to dominate their more timid brothers and sisters by pecking at them. Once they draw blood, other members of the same flock often pick upon the poor victim, until the chicken is reduced to a bloody mass of flesh and feathers. At this point, the mutilated bird should be removed from the flock and put out of its misery.

I have used anti-pick lotions and sprays in order to prevent cannibalism, but they have never worked. The ointments were usually red in color and acted like a magnet to draw the attention of aggressive birds.

Debeaking is one of the favorite methods employed to prevent cannibilism. The process of debeaking consists of removing about half of the upper beak and a small portion of the lower jaw.

Tiny rose-colored spectacles, called specs or peepers,
are attached to the bird's upper beak to prevent
cannibalism. Small nostril rings (below) make cannibalism almost impossible.

A debeaked bird cannot pull feathers, break the skin of another chick, or hook feed out of a trough. (The upper part of a chicken's beak is shaped like a hook. When chickens peck at food, there is a tendency for them to cast expensive feed out of the trough. This is a reason why some poultry raisers automatically debeak baby chicks. However, if you do not fill the feed trough more than half full, once the baby chicks have learned to feed from it, whether they have been debeaked or not is irrelevant.)

You can debeak a bird using a sturdy pair of nail clippers or wire cutters. The beak is likely to grow back within 9 to 10 days, at which time you can repeat the process. A precision electric debeaker costs about $75. It has an electrically heated blade that cuts and cauterizes the wound at the same time. You can also choose to have your chicks debeaked at the hatchery. However, the beaks will probably grow back in, and you will have to repeat the process.

There are other methods of stopping cannibalism. Tiny spectacles, called specs or peepers, can be attached to the top of the birds' beak, in order to hinder their vision. Some specs are rose-colored, in keeping with the belief that a chicken who views the world with rose-colored glasses is less aggressive. Small rings are also available, which, when fastened to the bird's nostrils, make it almost impossible for the bird to cannibalize other chickens. Specs and rings are used for mature birds only.

No one knows exactly why chickens cannibalize each other. The vice has been blamed on too much heat, boredom, overcrowding, incorrect feeding, lack of water, and too much light. The problem is not as severe when chickens are given access to a yard, rather than kept in total confinement. This should tell us something.

THE CHICKEN YARD

At 6 weeks of age the chicks are good size and are well-feathered out. At this point, I find the chicks benefit if they have access to an outside range. They will get exercise, sunlight, green feed, and a chance to vary their diet with bugs, slugs, grasshoppers, and minerals in the soil.

The first requirement of a good chicken yard is that it be secure from dogs, foxes, and other predators. The yard should be fenced as described on page 36–37.

In very hot and sunny weather, the chickens need shade. Any shrubs or trees enclosed within the yard can provide shade. Otherwise, you will have to erect small canopies or simple lean-tos for chicks seeking relief from the hot sun.

An ideal setup is to have the yard fence attached to the chicken house. With a small trap door at ground level in the chicken house, the chicks can go in or out as they choose. This will also save the expense of building a separate range shelter. You can leave the waterer and feed trough inside their house and put the flock inside at night for security.

It is very important to always use a clean range for the young birds. Do not use a yard that has been occupied by other fowl within the past year. Also, don't let any trash or wood piles accumulate around the chicken house or yard. They make good hiding places for rats and mice. Even if the chicken feed is stored in a tight container, the rodents still have access to the birds' waterer and feeder and can spread disease.

Advantages of a Yard

The advantages of a yard are many. First, it provides sunshine, exercise, and a more varied diet for the chicks. Bugs are chock full of protein. Admittedly, a small yard for 25 broiler-fryers isn't going to make much of a dent in your overall feed expense. The main advantage is that chickens with access to a yard always appear more bright-eyed and vigorous than their caged brothers and sisters. And, of course, they are less inclined toward cannibalism.

Disadvantages of a Yard

Cost of fencing and the cost of replacing chickens lost to predators are the main disadvantages of having a chicken yard.

If you can use your chicken yard year-round, it is easier to justify the expense. However, in cold northern climates, the yard will be utilized only in the summer.

Yards also increase the likelihood of the birds being exposed to internal parasites. One way to avoid this is to construct portable yards and shelters and to move the yard at least once a year. Moving the yard even more frequently so that the ground does not become pecked bare and muddy reduces the risk of exposing chickens to disease. A laying flock, which you keep a year or more, is more susceptible to worms than broiler-fryers that run in the yard for only a few weeks.

Predators are a major concern. In ancient times, white-feathered chickens were not popular with farmers because they were easy victims of birds of prey. Chickens of color, because their varied patterns of plumage provided a natural camouflage, were favored.

When chickens are on range, they are exposed to predators such as hawks, owls, foxes, coyotes, weasels, raccoons, skunks, and dogs. With the exception of the hawk, all the rest of the natural predators are usually nocturnal hunters. If you shut the chicks in a secure house every night, they should be safe from everything except hawks.

The white-feathered Cornish-Rock chick is a good target for hawks. The only way to control hawks is to shoot them or use baited traps set on top of fence posts, if local game laws allow it. I don't believe in killing hawks compulsorily. I think that an alert, active 6-week-old bird can run for cover and escape a falcon's talons, especially if confined to a small yard. It is assumed that the family dog and any strays will be kept under control.

FERTILIZER: AN ADDED BENEFIT

As an added benefit to raising broiler-fryers, count on your flock of 25 chicks to produce 100 pounds of manure, litter included, over the 8-week period. If the birds are on the range, the usable manure will be considerably less, but the yard will be well-fertilized – another reason to build portable yards and shelters and to rotate the yard regularly.

Poultry manure is very high in nitrogen. It is also very concentrated and has a lower moisture content than most other manures. Fresh manure from the poultry house is considered *hot;* that is, it will burn the roots of plants that come in contact with it. However, it can be collected and aged for a year, then spread directly on the lawn or garden as a fertilizer. Or it can be added to your compost pile, litter included.

SUPPLEMENTAL FEEDING

Farmers who are raising roasters and who will keep their birds for some 20 weeks in order for them to reach a weight of up to 8 pounds often feed whole grains during the late stage of the growth period. Corn is the most popular grain used as it is very high in energy. When the roasters are about 6 weeks old, they are given a small portion of corn along with the regular finishing ration. The amount of corn fed is slowly increased until about the fourteenth week, when the roasters are receiving about 50 percent corn and 50 percent finisher rations in their daily feed. To finish means, among other things, to lay on a bit of fat along with the meat, providing a more ample, filled-out, and tender carcass. Economics also play a part in this; corn is cheaper than commercially mixed finishing rations.

If you are aiming for a 4 pound broiler-fryer in 8 weeks, whole grains are unnecessary and could actually slow down the rate of gain of your birds.

If time is not of the essence, and if you have grain on hand, you could certainly feed corn along with the grower-finisher ration during the last couple of weeks of the broilers' growth period. Introduce the corn gradually, a few pounds per day, and then build up. It does not make sense economically to go out and buy a sack of corn for broiler finishing purposes at this time.

If you do feed whole grains, you must supply grit so that the birds can digest the hard kernels. Grit consists of tiny pieces of insoluble stone or gravel. A chicken has no teeth, and the presence of grit in its gizzard helps the bird in grinding food into digestible particles. Grit is not necessary for broiler-fryers if they are not fed whole grains.

The chickens will relish any table scraps, greens from your garden, fruit peelings, vegetable tops, and grass clippings that you care to feed them. They also like milk. If you give them milk, make sure the container you serve it in is made of plastic or enamel, not metal.

None of the above will cause any great surge in their weight gain, or take the place of the grower-finisher ration they are eating, but it will certainly spice up their diet. If you do feed milk, garden greens, or grass clippings, limit the amount to what the birds can eat in about 10 or 15 minutes. If they fill up on these supplements, it can hinder their intake of the grower-finisher ration, alter the balance of other necessary nutrients, and set back their growth rate. Always feed fresh and tender greens, as dried-out fiber can cause digestive problems in the birds.

BUTCHERING

When the chicks are 8 weeks old, check to see if they have reached a good slaughter weight. Catch 3 or 4 birds and weigh them on a hanging scale. It is easiest to catch them when they are in the house where you can shoo them into a corner. It is even easier if it is dark inside the coop and you use a flashlight. Try to catch the chickens by the legs, not the wings. And try to prevent bruising which will discolor the skin.

If the broilers weigh less than 4 pounds, plan to keep them for another week. If they average 4 pounds, you are right on target.

Plan to slaughter and dress at least 2 birds in your initial effort. If the first carcass comes out a bit uneven and ragged, don't worry. The second one will look much better. By the time you've done up to a dozen, you will be a real pro.

Catch the birds about 12 hours before they will be killed and isolate them in clean cages or pens. You can use orange crates, apple boxes, or any type of cage. It is best if the holding pens have openings on the bottom, such as wire mesh or wooden slats. This will allow manure to fall through the bottom and keep the bird cleaner.

Make sure the chickens are cool and comfortable. They will bleed out better if they are not overheated. Give them plenty of water, but no feed. This will make the evisceration process less messy.

If you have never killed or dressed a chicken before, ask an experienced poultry farmer to show you how and go through the whole process, step by step.

Necessary Equipment

You will need an axe or a very sharp knife, or both. You will need a length of cord or twine with which to tie the feet of the chickens. Or you can buy a killing cone. The bird is placed upside down in this holding device, with its head sticking out the smaller end. The cone restrains the bird from struggling and exposes its head and neck.

You will also need a bucket or plastic bag to catch the blood. A large metal container, such as a 10-gallon garbage can, is necessary to hold hot water for scalding the bird before plucking its feathers. A sturdy floating thermometer is helpful.

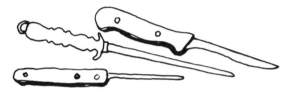

Sharp boning knives, kept sharp by a steel, make butchering and dressing go easier.

The chicken is placed upside down in the killing cone, with its head sticking out of the bottom. The cone restrains the chicken.

Killing the Bird

If you are going to kill the bird the old-fashioned way, with an axe, hold the chicken by both legs, upside down, over a block of wood, lower it until its head and neck rest lightly on the block, and cut off its head. Then hold the bird over the bucket or plastic bag until the struggling stops and the blood ceases to flow. It is best to let the bird bleed out thoroughly, as this makes for a better carcass.

If you are going to kill the bird with a knife, be sure it's a sharp one. A 6-inch boning knife works particularly well. Tie the chicken's feet with the twine and suspend it from the ceiling of the room or a rafter. Adjust the height so it is comfortable for you. Hold the head firmly with one hand, and sever the jugular vein that runs down the bird's neck. You can do this by inserting the knife into the neck close to the neckbone and turning the knife out. Or you can sever it by cutting from the outside. Cut the vein as close to the head as possible.

If possible, keep holding the bird's head over the bucket, until it has stopped struggling and is bled out. If you let go, it will splatter you and everything else in the surrounding area with blood.

Scalding

When the bleeding is over and the bird is limp, fill a large container or 10-gallon garbage can with hot water. The temperature should be 130 to 140 degrees F. and no hotter. Immerse the bird in the hot water and slosh it up and down for about 30 seconds. This up and down movement helps the hot water penetrate through the feathers and reach the skin, which it softens, thus making plucking easier. Keep in mind that if the temperature of the water is too hot, or if the bird is held under for too long, the surface of the body will begin to cook. You will notice that the skin becomes discolored and sometimes breaks open when the feathers are pulled out.

Picking the Feathers

Lay the bird down on a bench or table and begin plucking the feathers. You can start anywhere although some people prefer to begin with the large wing feathers. Do not, however, go against the grain. Always pull the feathers out in the direction in which they are growing. Otherwise, the skin may tear.

When you have finished picking all the feathers, there will be some tiny hairs left on the bird that you can't pull out. These hairs can be burned off by holding the carcass over the open flame of a gas stove, or singeing them off with a propane torch, candle, or, even, a cigarette lighter. If there are pinfeathers remaining on the carcass, use your thumb or a dull knife to press and squeeze them out.

The next step is to briefly wash the bird in cold water. Then cut its feet off at the hock, or first joint.

Eviscerating the Chicken

Now that the bird has been washed off, you can dress it out immediately. Some people let the carcass cool before cutting it up as it is easier to cut meat when it is a bit firm and resistant to the knife, not wigglely.

There are many ways to dress a bird. The following is the method I use and it has worked well for me over the years.

Place the bird down on its front side on a clean, hard surface, such as a counter top or table. With a sharp knife or scalpel, cut the skin on the back of its neck all the way down to the body. Peel the skin away from the neck and then turn the chicken over onto its back and pull the skin away from the throat and windpipe. With the skin removed, you can separate the throat (gullet) and windpipe from its attachment to the neck.

Follow the throat down to a pouch-like sack, which is called the crop (or craw) and is an enlargement of the gullet that serves to store food temporarily. Sever the gullet just below the crop and remove crop, windpipe, and upper gullet. Then cut the neck off as close to the body as possible.

With the bird still on its back, make a shallow incision, about 3 inches long, in the center of the soft area from the breastbone towards the tail, being careful not to go too far below the surface of the skin. You do not want to cut into the intestines which are just below the skin. Enlarge the incision that you have just made by pulling it apart with your fingers, breaking thin membranes, until the viscera is exposed and the opening is large enough to accommodate your whole hand. Put your hand inside the opening and work your fingers gently but firmly around the back of the mass of viscera, which will include the lower gullet, stomach, gizzard, and intestines. Pull the whole mass apart from the tissues that hold it in place. Then move to the rear end of the bird and make a circular cut around the vent, being careful not to cut too close and break into the intestine that is connected to it.

Pull the whole mass out through the front opening. The heart and liver should come out along with all the rest. Rinse the heart off and place it in a container of cool water. Before you do anything with the liver, remove the small, green sack, called the gall bladder, that is attached to it. You can cut or pinch the gall off of the liver. *Caution*: Do not break the gall bladder; the liquid inside can spoil any meat that it touches. After removing the gall, wash the liver off and place it in the container with the heart.

The next step is to open up the gizzard, by cutting around the edge of the muscle, until a yellowish area is exposed. Stop! Don't go any further. Pull the gizzard apart with your fingers and remove the inner sack, which contains food particles, without spilling its contents. Rinse the gizzard in water and place it in the container with the heart and liver.

Reach up into the body cavity and pull the pink and spongy lungs free from their attachment to the ribs.

The sexual organs are fastened to the backbone. They look like 2 little white beans and are easily removed.

Turn the bird over onto its front. With a deep triangular cut, remove the oil gland, which is located just above the tail.

Discard the intestines and all other offal either in a disposal pit or buried deep enough in the ground so that a hungry dog can't dig up a ripe feast.

Cooling the Carcass

After the bird has been drawn, it is important that the carcass be chilled immediately to prevent the growth of bacteria. Put the bird in a pail of ice water as soon as possible in order to lower the body heat. Try to keep the temperature of the cooling water from 34 to 38 degrees F., until the internal temperature of the carcass is less than 40 degrees F. Keep a close watch on it and don't spare the ice cubes.

Depending on the bird's size, it can take from 8 to 10 hours to cool a whole bird. If you prefer, you can cut the chicken into smaller pieces before cooling. Smaller-sized chicken parts will cool off in about 4 hours.

It is recommended that poultry be chilled for at least 12 hours before eating or freezing. Even a short aging period will make for more, tender meat. You can leave the bird in the ice water or else put it on a shelf in the refrigerator, but not in the freezer compartment, for 12 hours.

Before freezing the bird, let it drain for about a half hour until all the excess water is gone. A dry bird keeps better than a moist one.

Although you can cut the chicken into parts to speed the chilling process, you will find it is easier to cut the chicken after it has cooled. When cold, the body is firm and doesn't shy away from the knife.

If you want to split the bird, use poultry shears and cut along both sides of the backbone. Cut through the remaining skin on the front side. Separate the carcass by pulling it apart. You can quarter it by cutting between thigh and breast on both sides. This way you would have 2 pieces of breast and wing together and 2 pieces of thigh and drumstick combined.

PREPARATION FOR FREEZING

When you wrap any meat for the freezer, the main idea is to freeze the meat airtight. Moisture-proof materials, such as plastic wrap or laminated freezer paper, work much better than waxed paper. A small plastic freezer bag can hold the giblets. You can use rubber bands or freezer tape to bind the bag.

Cut the skin on the back of its neck all the way down to the body.

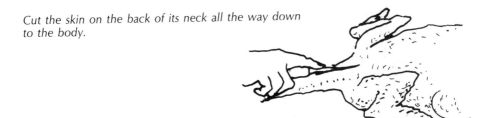

Pull out the crop and windpipe from the neck cavity. Then cut off the neck as close to the body as possible.

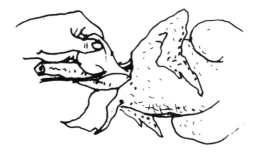

Cut around the vent, taking care not to cut into the intestine.

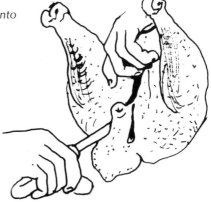

Remove the gall bladder that is attached to the liver.

If you use freezer paper, wrap it around the meat as tightly as possible, until the wrapping conforms to the shape of the portion being wrapped. If you use plastic freezer bags, try to fit the bag to the size of the chicken part being frozen and wrap it tightly. You can remove air from the package by inserting a drinking straw inside the neck of the bag and sucking out the air to cause a partial vacuum. Then twist the wrapper and tie securely.

Do not stuff a whole bird before freezing. This can cause the whole package to spoil because of slow temperature changes in the stuffing when the bird is being frozen.

For best results, flash freeze the chicken at -10 degrees F. and store at 0 degrees F.

How Long Can You Keep Frozen Poultry?

For giblets, the maximum recommended freezer storage time is about 3 months. Cut-up parts are good for about 6 months. A whole chicken will keep for 12 months.

FINAL THOUGHTS

When raising mixed sexes of chickens for broilers, do not entertain the idea of sparing the females from the axe in the hope that they will supply you with an ample amount of eggs. They will eat you out of house and home. If you are lucky, they may turn out 100 to 120 eggs in a laying cycle. This is about half of the production of a hen of an egg-laying breed. It just isn't economical to keep a bird of a broiler breed around for its eggs.

Chapter 4

RAISING
CHICKENS
FOR EGGS

Compared to raising broiler-fryers, raising laying chickens presents a far greater challenge. You can plan to have the laying chickens around for at least 17 months, compared to the 2 months involved in raising broiler-fryers. Over the course of those 17 months, you will be much more concerned with the birds' productivity, with disease prevention, and with health than you were with your broiler flock.

Housing needs are more elaborate, too. Layers require roosts, nests, and perhaps a dropping board or pit for the manure. You will have to provide warm housing during the winter, as well as artifical lighting for the laying birds during the fall and winter.

You will have to delay the sexual maturity of your laying birds. A pullet that comes into egg production too soon generally produces lots of small eggs; in doing so, the yound bird's reproductive organs may be harmed.

Disease prevention is an important concern. The birds will have to be vaccinated for the prevention of particular diseases and treated for internal and external parasites. You will have to cull hens that are not laying, experience the natural process of the birds molting, and perhaps discourage the occasional hen from becoming broody (meaning the hen ceases to lay new eggs and wants to hatch the ones she has laid).

All of this requires careful management. But you can look toward great rewards. Properly managed, the laying birds should be able to produce eggs year-round. A good laying hen will lay about 20 dozen eggs during a laying cycle, which will last about 12 months. Many breeds of laying hens make good roasters after their productivity as egg-layers has come to an end.

Lastly, a laying hen will produce about 25 pounds of manure a year. If allowed to age, it makes excellent fertilizer for the garden.

CLEARING UP SOME MYTHS

To begin with, 2 things should be made clear. One: you don't need to keep a rooster in order for hens to lay eggs. When she is born, the female baby chick has a fully formed ovary which contains thousands of tiny ova, or future eggs. When she reaches sexual maturity, the hen will lay eggs without a rooster egging her on, so to speak. The eggs will not, of course, be fertile. If you want fertile eggs, you must have a rooster to breed the hens. Fertile eggs are no more nutritious than nonfertile eggs and don't keep as well.

Two: white chickens don't necessarily lay white eggs, and chickens of color don't necessarily lay brown eggs. The color of the egg shell depends on the origin, or class, of the chickens and the color of their earlobes. Chickens that originated from the Mediterranean region and have white earlobes lay white eggs. Chickens of Asiatic, American, or English origin have red earlobes and lay brown-shelled eggs.

This difference in egg color is probably due to mutations occurring during the evolution of the chicken. Egg shell color involves multiple genes, and no one color is dominant over the other. A white egg breed chicken mated with a brown egg breed chicken will result in a hen that lays tinted eggs.

Contrary to old myths, the color of the shell has absolutely nothing to do with the food value of the egg. White or brown, the nutrients are the same.

SELECTING YOUR CHICKENS

Which breed should you select for your laying flock? There are a lot of choices here. The first thing to consider is whether or not you want to raise a lightweight chicken for eggs alone. When a lightweight hen's peak laying days are over, she is only good for soup or stew. On the other hand, under good management, this type of egg layer can produce a dozen eggs on about 4 pounds of feed, and she reaches sexual maturity at about 5 months of age, earlier than the heavier breeds. A pullet or young female, is considered sexually mature when she lays her first egg. The White Leghorn and other chickens of the Mediterranean class fit this category and are prolific layers of white eggs.

Or you can raise medium-size, dual-purpose chickens. These usually reach sexual maturity at 5½ to 6 months of age, eat more feed, and yield fewer eggs (although there are exceptions). These hens lay brown eggs. They make fine roasters when their most productive days are past. The American and some English classes of poultry are good examples of this type of chicken.

If you want chickens primarily for eggs, don't even consider raising the very large and heavy types like Brahmas, Cochins, Jersey Giants, and White Cornish. Although they are motherly, good setters, and often broody, they eat a great deal and their egg production is low.

Examine Table 4–1 which compares the characteristics of various egg-laying breeds. This should help you make your decision on which breed(s) to raise.

STARTING THE LAYING FLOCK

After you have decided upon a breed or variety of egg-laying chickens to raise, you have another very important decision from 5 choices. How do you want to start with your laying flock?

- **Incubate the eggs?**
- **Buy day-old chicks?**
- **Buy 8-week-old started chicks?**
- **Buy 20-week-old ready-to-lay pullets?**
- **Buy 2-year old hens?**

I suggest that you select the age or development that best fits your own personal timetable. But, as you will read, I have my definite preferences.

Let's look at the advantages and disadvantages in each category.

Incubating Eggs

Buying eggs to incubate is not reasonable. To begin with, you will have to buy about 4 dozen expensive hatching eggs, and put them under broody hens or in an artificial incubator for 21 days. If 60 percent of the eggs hatch (the national average), you will have about 29 live chicks, of which 50 percent, or perhaps 15, will be females and future egg layers. Then the chicks still have to be brooded for at least 6 weeks. Incubating eggs is fun, but it can be a long and costly route to go.

Day-Old Chicks

Day-old baby chicks, sexed (meaning all pullets, or female), of the best egg-laying breeds vary in cost from $.90 to $1.00 each. Thus, 15 pullets will cost about $15.00. If they are of the Mediterranean class and lay white eggs, they will start laying when they are about 5 months old. If the pullets are American class, dual-purpose birds,

Table 4-1

A SAMPLING OF THE CHARACTERISTICS OF CERTAIN BREEDS & VARIETIES OF CHICKENS

Breed or Variety	Origin or Class	Egg Color	Average Mature Hen Body Weight (Pounds)	Rate of Lay
Standard Breeds				
White Leghorn	Mediterranean	White	4.5	Excellent
Black Minorca	Mediterranean	White	4.5	Very good
Ancona	Mediterranean	White	4+	Very good
Blue Andalusian	Mediterranean	White	4+	Very good
Barred Plymouth-Rock	American	Brown	6–7	Very good
White Plymouth-Rock	American	Brown	6–7.5	Very good
Rhode Island Red	American	Brown	6+	Very good
New Hampshire	American	Brown	6.5	Very good
Silver Laced-Wyandotte	American	Brown	6.5	Good
Jersey Giant	American	Brown	7	Fair
Hybrids				
California Grey	American	White	4+	Excellent
California White	American	White	4.5	Excellent
Golden Comet	American	Brown	4	Excellent
Black Australorp	English	Brown	6–7	Very good
Buff Orpington	English	Brown	6	Good
White Cornish	English	Brown	8	Poor
Buff Cochin	Asiatic	Brown	8.5	Poor
Light Brahma	Asiatic	Brown	9	Fair
Araucana	South America	Tinted green & blue pastel	5.5	Good

Table 4–1. Choose the egg-laying breed that suits your needs.

and lay brown eggs, they will start laying at 5½ to 6 months of age. You will have to brood them for 6 weeks and should still expect to lose a couple of chicks before they are fully grown.

Eight-Week-Old Pullets

Buying 8-week-old started pullets is a good deal. At this age, they do not have to be brooded or supplied with artificial heat as they are well feathered-out. Although there are some hatcheries that will sell and ship started pullets, the mailing cost is high, considering the birds' weight. You will probably have to buy started pullets at a local hatchery within driving distance of your home. The mortality rate will be lower than that for day-old chicks, although you can still expect to lose a couple of pullets within a 12-month period. Fifteen started pullets will cost at least $2.00 apiece. I think they are certainly worth the price.

Twenty-Week-Old Pullets

Buying 20-week-old, ready-to-lay pullets can be expensive. Because someone else fed and cared for them, these birds will cost $4.00 to $5.00 each. However, their mortality rate is quite low. They can be purchased only from nearby hatcheries or from poultry farmers who specialize in this type of bird.

Two-Year-Old Hens

Two-year-old hens of lightweight breeds can sometimes be obtained from farmers in your locality who want to dispose of aging hens and start fresh with a new flock. They are a bargain at about $1.00 each. Heavier hens cost more. During their second laying cycle, these hens will lay 10 to 25 percent fewer eggs than they did in their first cycle. White Leghorns that laid about 240 eggs in their first year, may lay about 200 eggs in their second year. The heavier and medium-size breeds will lay proportionally fewer eggs.

Although it's safer and you can hedge your bets by buying 20-week-old ready-to-lay pullets or aged hens I don't recommend it. I look on buying older pullets and hens kind of like going to the ring-toss booth at the county fair and purchasing a Kewpie Doll from the carnie operator, instead of testing your own skill.

How Many and What Kind to Buy

If you want to keep chickens for eggs only, then I suggest you buy 15, day-old chicks of a lightweight breed. Just don't keep Leghorns or other strictly laying types with any hope of hatching out their eggs in subsequent years. These hens have become egg-laying machines, and any motherly traits or broody qualities have pretty well been bred out of them.

My own personal suggestion for the beginner is to buy 15, day-old pullet chicks of a dual-purpose breed, such as Barred Plymouth Rock, White Plymouth Rock, Rhode Island Red, New Hampshire, or Black Australorp.

In regard to egg-laying ability, it should be noted here that in national egg-laying contests, some strains of White Rock, Barred Rock, Rhode Island Red, and New Hampshire Hens have consistently laid over 300 eggs in 365 days. And, in an Australian program, a Black Australorp hen created a sensation when she laid 364 eggs in 365 days.

It must be said that this kind of egg production should not be expected of the rank and file chicks you purchase from a hatchery. But their outlay should be close to that expected of strict egg-layers. And as an extra bonus, the dual-purpose hen can give you a lovely roast when her duty is done.

It may be difficult to buy just 15, day-old, pullet chicks from a distant hatchery, as many of them will ship only a minimum of 25 chicks via Air Parcel Post. The reason for this is that it takes about 25 chicks in a packing box to keep each other warm and ensure safe arrival during a long journey. You can purchase 25 or 30 chicks, and split the order with a friend or neighbor. Otherwise you may have to buy your chicks locally.

I specify the number 15, as, everything being relative, and assuming a natural mortality loss of 2 or 3 chicks during their growing-up period, keeping any more hens usually will result in a surplus of eggs during their peak laying cycle. Although you can sell, trade, barter, freeze and use eggs in baking, it is often difficult to dispose of a surplus of eggs. (See chapter 5 for suggestions on keeping and freezing eggs.)

Another alternative is to buy 25 straight-run nonsexed day-old chicks of a medium-size dual-purpose breed. Half of these chicks will be females, which you can raise as layers. The other half, the males, can be raised as broiler-fryers or roasters. Straight-run chicks cost about half as much as sexed pullets.

However, the mixed sexes do create a problem. After about 6 weeks, the pullets and cockerels (males) should be separated. Their feed requirements differ at this stage, and the pullets will need nests and roosts eventually, neither of which are necessary for the cockerels. Thus, it will be necessary to maintain 2 different pens or coops for the birds after 6 weeks: one for the layers and one for the meat birds.

When to Buy Your Chickens

If you are going to incubate eggs, you should start them in April, so you can brood them when the weather is warm. Started pullets, if available, can probably be purchased in midsummer. They should be ordered about 3 months before you need them. Ready-to-lay pullets will most likely be available in September or October. They should be ordered about 6 months ahead of time. Aged or 2-year-old hens from a farmer who wants to replace an entire flock are most readily available in late spring or summer.

If you follow my advice and go for day-old chicks, the best time to buy them is in the spring, preferably the month of May. You will brood them in the warm weather and they will then start laying in October. Order them a month ahead of time; that is put your order in in April for a May delivery.

Where to Buy Your Chicks

The same rules that govern buying chicks for meat production apply here. Order from a reputable hatchery and buy the best quality chicks that you can. Bargain chicks usually are not bargains at all.

Your potential laying stock should come from a hatchery that is classified as "US Pullorum-Typhoid clean." The chicks should be the offspring of proven birds that have been developed for a high rate of production. The USDA publishes an annual report on production tests of various breeds and hatcheries. Copies of this publication can be obtained by writing to the Poultry Research Service, USDA, Beltsville, Maryland. Even better, ask local poultry farmers and your extension poultry specialist for advice on where to buy your chicks. It's best to buy chicks locally. Long-distance transportation isn't the best way to get a flock started although the chicks will come with guarantees, and dead chicks will be replaced or your money refunded – if you deal with reputable hatcheries.

HOUSING

Before your chicks arrive, you will have to get their housing in order. The housing requirements for a small flock of day-old pullets are the same as for broiler-fryers – at least for the first 6 to 8 weeks of their lives. They will need a clean, warm pen, draft-free, and secure from predators. After that, you will have to provide roosts, nests, and a dropping board and pit for manure collection.

The Brooding Pen

I recommend that you raise your laying flock on the floor with litter. You can use a chicken coop or partition off a small area in a barn or shed. The pen must be well cleaned, disinfected, and aired for 2 to 4 weeks before the chicks arrive.

In the beginning, each chick will need about .5 square feet, if raised on litter on the floor. By the time they are 6 weeks old, the chicks will require 1 square foot of space each. When they reach laying age, around 5 months, each pullet will need 3 to 4 square feet, depending on whether they are lightweight or medium-size birds. The more room they have, the better off they will be.

First, though, you will keep your chicks within a chick guard, as described on page 43. The guard will create a chick corral, about 6 feet in diameter, to keep the chicks close to the source of heat, water, and feed.

Heat

If you raise the chicks on the floor, you can use an electric or gas-fired hover to keep them warm. The hover provides a heat source and a canopy, under which the chicks can gather. The canopy rests about 1 foot over the floor, and the chicks can take advantage of the heat, or escape it, as necessity demands. Or you can use a 250-watt infrared heat lamp as the heat source. The heat lamp setup should include a porcelain socket, a reflector on top, and a metal guard around the bulb.

The temperature should be maintained at about 95 degrees F. during the first week of brooding. (See pages 38–40 for more information on setting up the heat lamp and adjusting the temperatures.) As the chicks grow, plan to reduce the temperature in their pen by about 5 degrees per week, as long as the chicks seem comfortable with the change. By the time the chicks are feathered out, the pen temperature should be about 75 degrees F. At this point, the indoor and outdoor temperature should be equal, and you can discontinue the heat. However, if nights are cool, be sure to turn the heat back on.

Litter

The purpose of the litter is to provide warmth and to absorb moisture. Your choice of litter will depend on what is available to you. Wood shavings, sawdust, ground corncobs, rice hulls, peanut hulls are often used. Start with 3 to 4 inches of litter.

No Cages, Please!

I do not like cages. To begin with, if you are going to use cages, you must have central heating for the whole chicken house. You can't train a heat lamp on a small cage; the chicks will have no place to go to cool off.

Also, when birds reach laying age, those raised in cages are subject to cage fatigue, a paralytic condition most comon among high-producing young pullets during summer months. Also the disease called "fatty liver syndrome" is most commonly seen in caged birds. This results in excess quantities of fat deposited in the birds' liver and other body cavities.

I also think that cages promote cannibalism. After all, when a hen has eaten her fill of that same old feed ration coming down that automated conveyor belt, and has drunk enough water from the continuous flow system in the cage, and laid her egg for the day and as soon as she dropped it, the egg rolled quickly out of sight into a collecting trough, and it wasn't even around long enough for her to cluck over it for a few minutes . . . . Well, c'mon, what else is there for her to do, if she's confined with 2 other hens, in a standard wire cage, 12 inches wide and 16 inches deep, but to pick on them?

Ventilation

In the summertime, the chicken house should be cooled by ventilation. This is easy if there are windows on more than 1 side of the house and cross ventilation can be provided. You are not looking to create drafts, especially on the floor, just a good exchange of air movement to help keep the litter dry.

Ventilation removes moisture from the chicken coop and prevents high humidity. When it is very hot, high humidity causes eggs to deteriorate rapidly and become moldy. Even in cold weather, ventilation is important. It is best in cold weather if only the windows on 1 side of the house are opened. If possible, open the windows with the southern exposure. The main idea is for an exchange of fresh air for foul, no pun intended. It is better to have a cool, dry chicken house than a warm, wet one.

EQUIPMENT AND FURNISHINGS

When you consider the equipment for laying hens, remember that caring for these birds is a long-term commitment. Automatic waterers will reduce your labor considerably, for example. In addition to waterers and feeders, which are required for broiler-fryers too, laying pullets should be furnished with roosts and nests. A manure dropping board or manure pit will keep the chicken house clean.

Waterers

A 1-gallon fountain waterer will be sufficient for 15 pullets from the time they are 1 day old until they are fully mature. If running water can be piped to the chicken house, consider investing in an automatic waterer, which will save you a lot of labor and guarantee that the birds have a constant supply of water. Alternatively, you can water birds from shallow pans, but they should be covered with a wire guard so that baby chicks cannot climb inside and drown. See page 41 for an illustration of a homemade waterer constructed from a 1-gallon can.

Ample water must be provided even in extremely cold weather. Keeping the water supply from freezing in northern zones can be done in several ways. You can purchase an electric immersible device that will keep the water from freezing in a bucket or container. The old-fashioned way, which works quite well, is to bring a container of warm water each morning and night to the chicken house, set it up, and take the frozen receptacle back to your kitchen to melt and use next time.

Feeders

Once they have learned to eat from a trough, each day-old bird will need only about 1 inch of feeding space. Adult hens need about 3 inches each. Thus, a single feeder, 18 inches long and accessible from both sides, will be sufficient for feeding the 15 pullets from day 1 until they are laying eggs.

In the beginning, of course, the trough will be a bit large for just 15 chicks, but as long as the edge is about on a level with the chicks' backs, they will be able to reach down and feed. By the time the pullets are mature, the feeder will be just right. There is no reason why you couldn't start out with a small trough and graduate to a larger one when the birds are grown. It is best if the trough has adjustable legs so that you can raise it as the birds grow in height. The feeder can be made of metal or wood.

Roosts

Although not absolutely necessary, laying hens seem to enjoy and favorably utilize roosts in their pens. A roost is simply a lodging place, or perch, for the birds to sit on when they sleep at night. Light-weight chickens of the Mediterranean class, such as Leghorns, most likely had ancestors that roosted in trees, thus their predilection for roosting is probably inherited.

Roosts are usually made of 2-inch stock lumber with the top edges rounded off or beveled, providing a natural contour for the shape of the chickens' feet as they hang on for dear life above ground in the dark. The rounded edge also helps prevent injury to the chicken when she settles down for the night. The roosts should be about 2 feet above the floor and placed at least 1 foot apart. If dropping boards are used, the roosts are mounted higher. Light breeds of chickens will need about 8 inches of space per hen, while the heavier types require 10 to 12 inches each. Five roosting poles, each 30 inches long and attached to the walls of the chicken house, will provide plenty of room for your 15 hens to roost at night, no matter how big they are.

It is recommended that if you raise young birds as layers, you should provide roosts in the house they are brooded in so they will become accustomed to roosting at night when they are fully grown.

One season I kept laying hens in a large, open barn that also housed cattle. That winter, weasels moved in and began to decimate my flock, one by one. Every other morning, I'd find another one of my good girls headless on the floor. Thereafter, each night I went to the barn and with a flashlight caught the remaining birds in the dark and placed them up on the rounded piping above the cattle stanchions to roost. I figured that if something disturbed my chickens in the night, a squawking hen thrashing around in the dark, sharp claws pointed down and landing on the back of 1200 pounds of nervous and pregnant Hereford cow, would scare off any predator. The system wasn't foolproof, but it did work for a while.

Dropping Boards

The purpose of a dropping board is to catch all the feces the birds drop as they roost. The board, or boards, are made of solid pieces of wood of dimensions wide enough to catch all the droppings. They should be mounted onto the wall about 2-½ feet above the floor. This leaves the floor space for the birds to move about, exercise, or do whatever they like, on the litter below them, without encountering manure. The boards should be cleaned at least once a week; they can be cleaned daily.

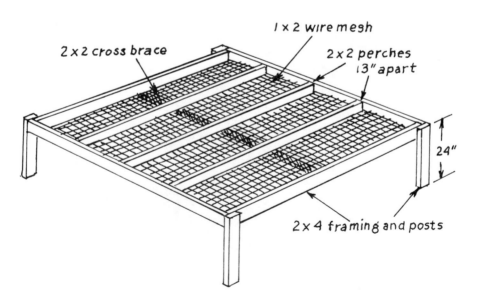

Some roosts are set directly above a dropping pit.

The Dropping Pit

A dropping pit is a more substantial fixture than a dropping board. It usually consists of a metal pan or wooden structure directly below the roosts, large enough to catch all the droppings. The pit will have sidewalls 18 to 24 inches high and a chicken wire covering so that the birds cannot get into it. This type of pit does not have to be cleaned as often as the board. Unfortunately, it also attracts flies and rats.

Some poultry raisers place feeders and waterers above dropping pits, in the center of the pen. This serves to keep the floor of the house cleaner and drier.

Nests

Laying hens must have nests to lay their eggs in so that the eggs aren't dropped at random on the floor, becoming dirty or broken. Eggs that aren't in nests promote egg-eating by other chickens, as it is the nature of the bird to peck at anything, once.

A nest should be just large enough to hold 1 hen. It is usually made of wood and is 12 inches wide, 12 inches high, and 12 inches deep. There should be a sloping roof over the nest with a pitch steep enough to discourage hens from roosting on it. There should also be a small perch at the opening so that the hen can stop for a minute, look within, inspect it carefully, and cluck a few times before going inside to do her thing.

The bottom of the nest should be covered with 3 or 4 inches of clean litter. There should be a 4-inch board at the opening, up front, in order to keep the bedding from falling out. Nests should be placed 18 to 24 inches above the floor. Provide 1 nest for every 4 or 5 hens. Four nests will provide plenty of space for your 15 layers.

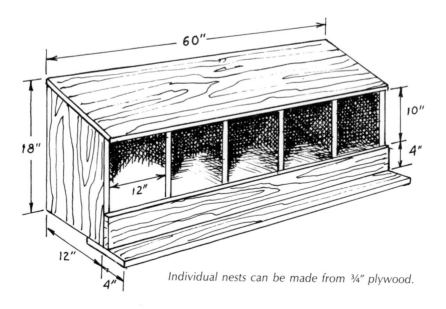

Individual nests can be made from ¾" plywood.

If you supply roosts, dropping boards, and a perch, or landing board, directly in front of the nest, the laying space itself will not be contaminated with droppings. Keep the nests as clean as possible, inspect them daily, remove soiled bedding, and add fresh litter.

Fake or glass eggs don't really seem to spur hens on in laying eggs. Also, it is the nature of chickens to monopolize 1 nest and totally ignore another perfectly good nest. Nests should be dark. Hang a heavy cloth or piece of canvas from the top of the nest to cover about two-thirds of the entrance to the nest to serve this purpose.

Nests can be purchased from agriculture supply houses. However, the design is quite simple, and they can be easily made from plywood or 3/4-inch lumber.

CHICKENS ON THE RANGE

It has been said that a chicken yard cannot really be too big. However, fencing is very expensive. I would rather say that any size yard, no matter how small, is better than no yard at all.

Although chickens can eat up to 15 percent of their nutritional requirements when pastured on legumes or certain grasses, it is not reasonable to plant a crop for just 15 birds.

Fresh air, sunshine, exercise, green feed, and eating insects can only improve the health and strength of chickens that have access to a yard. You must still feed them their regular rations, besides whatever goodies they find in the yard.

When large numbers of chickens are raised on range, some poultry producers feed rations in the form of pellets or crumbles, rather then mash, which are not subject to blowing away in windy weather. However, pellets and crumbles cost more than mash; and the birds seem to eat it faster, and thus have more time to get into trouble or become cannibalistic.

An outside yard or range should be well-drained, as standing water or mud holes can cause disease problems in your flock.

It is best if the fence of the yard is attached to the chicken house, with a small door cut in the side of the coop, so that the hens could go in or out at will. The feeders and waterers can be kept inside the house, providing an incentive for the birds to go inside, so you can shut them in at night, safe from predators, such as owls, coyotes, foxes, and all the other nocturnal hunters. Daytime predators include hawks, which will attack chickens, and crows and ravens, who will eat any eggs laid outside in the yard.

In very hot weather, the hens can go inside to escape the heat. If the yard is on the southern side of the house, the hens can even get some fresh air on a sunny winter day.

If you do provide a yard, don't let trash or old lumber piles accumulate, as they are perfect breeding grounds for mice and rats.

RATS AND MICE

As long as the subject of rats and mice has come up, we might as well deal with it now. These pesky rodents will be attracted to the chicken feed. You must store the feed in tight containers; metal garbage cans with lids work very well. Rats and mice not only eat a lot, but they can spoil up to 3 times as much feed as they eat. Anti-coagulant poisons are very effective in getting rid of rodents. Setting up a permanent bait station, with the poison placed along a path that you have reason to believe is customarily used by rats or mice, should help keep the rodent population under control. But be sure other animals, especially your pets, can't get into it.

BROODING THE CHICKS

If you begin with day-old chicks, the brooding procedure is the same for the little potential egg layers as it is for broiler-fryers—for the first 6 weeks of their lives (see pages 49–51). You must keep them warm, dry, well-fed, watered, and free from cold drafts and predators.

This small chicken house and totally enclosed range provide enough space for 8 hens and provide complete protection from predators. (Photo courtesy Joseph Blanchette)

Set up a routine for feeding, watering and observing the chicks. These chores should be done twice a day, every day. Make sure the chicks are warm enough. Cull any sick birds.

If you buy 8-week-old started pullets, ready-to-lay pullets, or older hens, brooding is not necessary. They will have grown their feathers and not need artificial heat.

FEEDING EGG-LAYING CHICKS

For the first 6 weeks of age, egg layers are fed the same starter ration that broiler-fryers are fed. After that, the young pullet is fed a ration lower in protein to delay her sexual maturity. This gives her organs a chance to become fully developed before she starts laying eggs.

Feed Program for Day-Old Pullet Chicks

For the first 6 weeks of their lives, you will feed pullets a starting ration mash, consisting of about 20 percent protein and containing a coccidiostat to prevent coccidiosis. The young chicks must have mash as their digestive system cannot handle whole kernels of grain at this stage of their development.

Keep the starting mash in front of the chicks at all times but only fill the trough half full in order to prevent wasting of feed. It is best *not* to put fresh feed over old feed. Arrange your feed purchases so that you don't store commercial feeds too long (1 month is okay), and always use tight containers for storage.

When you feed the birds in the morning, take care of their water supply at the same time.

Water

The chicks must have an ample supply of clean, fresh water available to them at all times. Rinse out the waterer before filling it up each time. If possible, the water should be lukewarm.

When the Chickens Are 6 Weeks Old

When the chicks are 6 weeks old, begin feeding them a growing ration of 16 to 17 percent protein, also containing a mild coccidiostat. The main idea here is to cut back on the protein level so that the pullets don't come into egg production too soon and possibly injure their reproductive system.

Always try to change feeds gradually. If you have starting ration left over, mix it in with the new growing ration.

If it is a problem to obtain a growing ration containing a coccidio-stat, the medication can be purchased from you local feed or farm store and added to the birds' water supply. When giving medication, always follow the manufacturer's instructions meticulously.

When the Chicks are 14 Weeks Old

When the chickens are 14 weeks old, withdraw the coccidiostat from the growing ration. It is believed that if chicks are treated with a coccidiostat until they are about 14 weeks of age, they will develop an immunity to the disease.

At 18 Weeks

When pullet chicks are 18 weeks old, some farmers begin feed-ing a laying mash of 15 percent protein. Others continue on with the growing ration until about 5 percent of the pullets have begun laying eggs, and then they switch to a laying mash. If you are only raising 15 pullets, the switch would come after the first couple of eggs are laid.

If you are puzzled by my making such a big deal about switching from a growing ration of 16 percent to a laying mash of 15 percent, a difference of only 1 percent protein, here is the reason. The laying mash contains 2 times more vitamin A, 2½ times more vitamin D, and at least 3 times more calcium than the growing ration. All of these nutrients are most important for hens laying eggs.

As the egg shell is almost completely made up of calcium car-bonate, a hen needs all the calcium she can get in order to develop a strong shell. Besides the calcium supplied in prepared feeds, poul-try farmers also provide it in the form of oyster shell, crushed lime-stone, or ground bone meal, all of which can be purchased at farm-feed stores.

These supplemental forms of calcium can be placed in separate boxes or trays, next to or close by the feed troughs, thus giving the hens a free choice for their needs. Do not feed extra calcium to the birds until they are laying.

HOW MUCH FEED IS EATEN
BEFORE THE FIRST EGG?

It takes 20 to 25 pounds of feed to develop a lightweight chicken, such as the Leghorn, to her sexual maturity or egg-laying stage, which occurs at about 20 weeks. Heavier or dual-purpose breeds, such as the Plymouth Rocks, New Hampshires, or Rhode Island Reds, will eat 10 to 15 percent more feed and take 2 to 4 weeks longer to lay their first egg.

HOW MUCH FEED DOES
A LAYING HEN EAT IN A YEAR?

A mature, lightweight bird, again using a Leghorn as an example, will eat about 1/4 pound of feed daily, or about 90 pounds of feed in a year, while producing an average of 240 eggs or more. The dual-purpose breeds will eat 10 to 15 percent more food.

Keep in mind that laying hens will eat more in the colder months of the year in order to help themselves maintain a suitable body temperature for survival. In the winter, some poultry farmers feed whole grains to their layers at the afternoon or evening feeding, as it takes a hen longer to digest hard kernels than the ground-up laying mash and the prolonged digestive process can help keep the bird a bit warmer on cold winter nights.

HEN PHYSIOLOGY

Due to her extraordinary digestive processes, a working (laying) hen can digest food in about 2½ hours. That is, from the time she picks it up with her beak and swallows the food until she is ready to excrete the waste as feces only 2½ hours elapse. I think this is remarkable. On the other hand, a loafing, nonproductive hen can take up to 12 hours to digest her food.

It takes a hen 24 to 25 hours to produce an egg. About 30 minutes after an egg is laid, another yolk is released from her ovary to be laid the following day.

FEEDING HOMEGROWN GRAINS

If you have homegrown grains, such as corn, wheat, barley or oats, or if they are easily available, you can use them to provide an economical supplement to expensive laying mash. Anything pre-mixed or packaged always costs more than the prime article.

Do not feed whole grains to chickens until they are 6 to 8 weeks old, as their digestive system cannot handle it.

If you are going to feed whole grains, start out with a ratio of about 10 pounds of mash to 1 pound of grain. Don't plunge blindly into formulating your own feed ration; ask your extension poultry specialist for advice in putting together a well-balanced adequate ration.

Corn is one of the main feeds used in a mix of grain and mash. Corn provides the pigment that colors the birds' skin and the yolk of the egg yellow. Corn is high in energy but lower in protein than wheat or barley.

Increase the proportion of grain very gradually. By the time the birds are about 20 weeks old and ready to start laying, the ratio of grain to laying mash can be about 50–50. Each lightweight layer will be eating 7 to 8 pounds of feed per month, and half of that feed can be whole grains. However, homegrown grains are usually deficient in protein. Since you are aiming for a ration of 15 percent protein for your layers, feed them a commercial laying mash of about 24 percent protein in order to bring a 50–50 ration up to the proper protein level. I repeat, consult an expert on this.

THE IMPORTANCE OF TRUE GRIT

Grit is small insoluble particles of stone or gravel that, after ingestion, lodge in the chicken's gizzard. The gizzard is a posterior stomach of the bird, and its thick walls are rippled and muscular. The action of the gizzard, along with gastric juices secreted by stomach glands, helps grind food down to digestible size. If fed anything other than fine mash, chickens must have grit.

Feeding Hints

Some poultrymen scatter whole grains in the litter of the hen house, thus providing the birds with useful exercise in scratching for food. The scratching action stirs up the litter, helping to keep it dry.

FEEDING STARTED PULLETS AND MATURE HENS

Eight-week-old started pullets should be fed a growing ration of 16 percent protein, plus coccidiostat, until they are about 14 weeks old. Then withdraw the coccidiostat and continue with the 16 percent protein growing ration until they are 20 weeks old (5 months). At 5 months of age, the pullets will begin to lay. Feed them a laying mash and supply extra calcium.

Twenty-week-old ready-to-lay pullets or mature hens should be fed a laying mash of 15 percent protein, plus extra calcium. Provide grit, if the birds are fed whole grains.

Protein requirements are highest when chickens are laying the most eggs. Large-scale commercial poultry producers practice phase feeding, which consists of feeding rations of various protein levels, to accommodate their hens' production of eggs. This procedure is quite involved and is not economical, unless you are feeding thousands of birds.

SUPPLEMENTAL FEEDING

Chickens relish greens from the vegetable garden and table scraps. It seems to perk up their appetite. Onions and some fruit peelings can impart a distasteful flavor to their eggs and should not be fed. Also, only feed an amount of this type of supplement that can be cleaned up by the birds in 15 or 20 minutes. If they fill up on kitchen scraps and garden greens, they will eat less of the carefully formulated, well-balanced laying mash ration, and you will get fewer eggs.

GATHERING EGGS

The fun part of raising layers is filling a basket with beautiful fresh eggs. Eggs should be gathered at least 2 times a day. In very hot weather, they should be gathered 3 or more times a day. Eggs exposed to high temperatures deteriorate rapidly. In the wintertime, eggs will freeze. Eggs left in the nest can become dirty, broken, or even tempt hens into becoming egg-eaters. For more on handling eggs, see chapter 5.

Gather eggs at least twice a day. (Photo courtesy Joseph Blanchette)

THE NATURAL EGG-LAYING CYCLE

A pullet's first eggs will be very small in size, and you can expect her to lay an egg every 3 or 4 days. It is believed that those pullets that lay eggs first will turn out to be the most productive hens in a flock. After 4 to 6 months of steady laying when she has reached her peak, a hen of a lightweight breed, such as a Leghorn, will be laying about 12 eggs in 16 days on about 4 pounds of feed. Unless of special egg-laying strains, the medium-weight or dual-purpose hens will be laying slightly fewer eggs and eating up to 5 pounds of feed per dozen eggs produced. Fifteen pullets will supply the average family of 4 with an ample quantity of eggs.

How Many Eggs Can You Expect?

Normal mortality rate for 15 chicks during the brooding and growing period is 1 chick. Average loss during a 12-month laying cycle for 14 birds is 1 hen. The balance: 13 live, laying hens.

In one year, 13 good hens will lay about 3,120 eggs. If, in a family of 4, each member eats 2 eggs per day, that totals up to 2,920 eggs. This still leaves a surplus of some 200 eggs that can be stored in your refrigerator, frozen, used for baking, fed to family pets, traded, bartered, or sold to friends and neighbors.

These figures are not hard and fast, just an example of what you may expect. Some members of a family probably don't want eggs every day, no matter how fresh they are. Others may desire eggs for Sunday breakfast only. Thus, during a normal egg-laying cycle, it is usually a case of feast or famine. In the beginning, there will not be enough eggs for the whole family. At the peak of the laying season, there will be a surplus of eggs. Toward the end of the hens' laying cycle, when production rates wind down, there may not be enough eggs for the whole family. It is a good idea to plan ahead as to how you will dispose of surplus eggs when your flock is laying at its peak.

MOLTING

When hens reach the age of 17 to 20 months, they begin to lose their feathers and stop producing eggs. This very natural periodic shedding of feathers is common to all birds. The process is called molting.

The molting process is triggered by secretions of the thyroid and pituitary glands. Chickens cast off their feathers, sometimes one by one, and sometimes in bunches. The molt can last from 2 to 4 months. Those hens that molt the quickest, that is, lose feathers rapidly and then grow new ones back in a short time, get back into egg production quickly and are the best layers.

It is a common practice for those commercial poultry producers who raise thousands of layers to sell off their flock when molting occurs and the first laying cycle ends. They do this because the hen will still eat about 6 or 7 pounds of expensive feed each month for several months and produce nothing.

Some farmers practice a program of forced or controlled molting. They bring their layers into an early artificially induced molt after the hens have been laying for about 9 months and are about 14 months old. At this point, the layers have passed their peak production. After a rest period of a couple of months, the birds start laying again, and the eggs are of better quality and production rates are higher than is usually the case if birds are left to their own natural molt. These farmers bring the birds into an artificial molt again when the birds are about 22 months old. They have found that they can squeeze more eggs out of a poor old biddy hen if she goes through 2 contrived molts and 2 egg-laying cycles within 22 months, than if she is left to her own natural mode — in which she would lay eggs for 1 cycle until she is 17 to 20 months old, and then go into a molt.

After a molt, and during their second egg-laying cycle, hens will lay larger but fewer eggs, and the quality of the eggs may decline.

Although I do not recommend the practice of forced molting for the beginning small flock owner, this is how it is done. The birds are confined in a dark or windowless chicken house. All lights are turned off, and they are fed whole grains instead of laying mash. If they do not begin molting within a short time, water is withheld from them for about 12 hours each day. After they start molting, regular feeding is resumed, and they are given all the water they can drink.

FACTORS AFFECTING EGG PRODUCTION

Hens that have access to a yard lay eggs with stronger shells. In fact, many environmental conditions have a great impact on egg quality and the egg-laying cycle. Here's an overview of some of the factors that may affect the number and quality of eggs your hens will produce.

Water

Chickens must have plenty of clean fresh water at all times. The body of the hen is largely made up of water, with averages ranging from 60 to 75 percent. Older birds have less and younger chicks have more water in their systems. The egg that the hen lays contains about 65 percent water.

Ample water must be provided at all times. In very hot weather, with temperatures above 80 degrees F., a hen can double her daily water intake. Take care to provide plenty of water, even in the winter. The birds need water, not ice!

Feed

The laying hen must be fed a ration containing all of the necessary nutrients that will allow her to maintain herself and also produce eggs. These nutrients include: proteins, carbohydrates, fats, vitamins, and minerals—especially calcium. They must be supplied in the correct amounts. Without adequate nutrition, a hen will lay a limited number of eggs or none at all. In some instances, as when an insufficient amount of calcium is supplied, the quality of the shell will suffer.

Lighting

If you want your pet canary to sing more, you put a night-light next to his cage. And if you want consistent, high egg production from your laying flock, you provide them with a minimum of 14 hours of light each day. Light activates the hen's pituitary gland, which in turn causes the secretion of hormones in her ovary, which stimulate her to lay more eggs.

Earlier in this chapter, I suggested that you begin your chicken-raising venture in late spring, in order to take advantage of temperate weather during the pullets' most vulnerable brooding period. If you start with day-old pullets in May, they will begin laying their first small eggs around October (5 months later). In the United States, particularly in northern zones, we do not get 14 hours of daylight in October. You will then have to supply artificial light in order to meet the requisite 14 hours of daily light.

One 40-watt incandescent bulb in a chicken house that measures 100 feet square or less will provide enough light to stimulate the hen and do the job. Provide a 60-watt bulb in a pen 200 feet square.

You can turn on the electric light before daybreak, or after sundown, or in any combination that suits you, *but*, it must be consistent and sustained, once you start the routine. Depending on where you live, with spring-born chicks, you will have to supply supplemental lighting from October to April or May of the following year. As soon as the natural daylight lasts 14 hours, artificial lights can be discontinued.

If you are raising your birds in a house without windows, the mathematics are easy. Turn the lights on for at least 14 hours, and then turn them off.

If your flock is already laying in late summer, whether they are in their first laying cycle or their second, commence with artificial lighting anywhere from mid-August to September, depending upon your geographic and climactic zone. Continue with the lighting until the following April or May in order to ensure the highest egg production from the flock. It is most important to be absolutely consistent with the lighting program. Hens subjected to irregular lighting will lay fewer eggs, or begin to molt and stop laying altogether.

Many poultry farmers use automatic timers to turn lights on and off, saving them the trouble of going to the hen house early and late. These systems should have a dimmer switch as well, so that the hens can find their favorite bedding place for the night. If you do it the hard way and go out to the chicken house at night to turn off the lights, if another electric outlet is available, it is a good idea to have a very low wattage bulb, say 7½ watts plugged in. After the brighter lights are turned off, leave the dim light on for a couple of minutes so that the chickens can find a place to sleep. It's no fun for a hen to try and find a skinny roosting pole located 2 feet up in the air when it's pitch dark.

Weather

Weather plays an important part in egg production. Hens lay best when the temperature is between 55 and 80 degrees F. Any temperature colder or warmer than that can cause a slump in the hens' output of eggs.

REPLACING HENS

It is common procedure for commercial egg producers to sell off their entire flock after the birds have been in production for a year or more, have passed their peak, and are due to molt and stop laying. It is most uneconomical to feed thousands of nonproducing birds for 2 months or more. These farmers replace the old hens with a brand new flock of pullets and start over again.

However, if your small flock produced well during their first laying cycle, it might be worthwhile to carry them through their molt and keep them for a second laying cycle. They will lay larger, but fewer eggs the second time around.

Culling Hens

On the other hand, it is just common sense to cull nonlaying chickens. You are simply separating the workers from the freeloaders. A nonlayer will eat just as much expensive feed as her laying sister, and produce nothing.

With some guidelines to go by, and a little experience, it's easy to separate the good layers from the nonproductive birds.

The laying hen is a totally dedicated mother and gives her all when producing eggs. If the egg shells need more calcium, she donates it without complaint from her very bones. She gives up her own natural coloration in order that her egg yolks come out yellow. She sacrifices the yellow coloration from her legs, beak, ear lobes, and eye ring.

Do not try to cull hens during a molt. You won't be able to distinguish the layers from the nonlayers. The best time to cull is during the flocks' peak laying season, and you should examine all of the parts of the body when culling.

Table 4-2

	AT A GLANCE: SEPARATING LAYERS FROM NONLAYERS	
Feature	**Laying Hen**	**Nonlaying Hen**
Comb	Large, red, waxy, full	Small, pale, scaly, shrunken
Wattles	Large, prominent	Small, contracted
Vent	Large, moist	Dry, puckered
Abdomen	Full, soft, velvety, pliable	Shallow or full of hard fat
Pubic bones	Flexible, wide open	Stiff, close together

Table 4-2. (From: Culling for High Egg Production, *Vermont Agricultural Extension Service)*

The comb and wattles (those hanging appendages on both sides of beak and upper throat) of a laying hen are large, bright red, waxy soft, and smooth. Her skin is soft and loose. The abdomen is full and pliable. The vent is enlarged, dilated, and moist. The pubic bones are wide apart, at least 3 to 4 inches. Her legs are bleached fairly white, and her feathers probably will be worn and soiled.

The comb of a nonlaying hen is dull in color and shriveled up. The skin is heavy and fatty. The abdomen is hard and nonpliable. The vent is dry and shrunken, and her pubic bones are hard and close together. Her legs are yellow, and her feathers nice and clean.

PREVENTION OF DISEASE AND PARASITES

As you will keep laying hens on your premises for a greater length of time than meat birds, 17 months or more, the chance of their contracting diseases and parasites is much greater.

To begin with, buy the best quality chicks that you can, of a parent stock from a hatchery participating in the National Poultry Improvement Program. No amount of good management can be successful with poor stock. Feed the correct feed and supply them with plenty of fresh clean water. Always follow proper sanitation practices in the chicken house where the chicks are growing up.

That's half the battle right there: quality chicks, the right feed, clean water, and good sanitation. The other half is observation and prevention.

When you go out to the chicken house to feed, water, and clean up the place, don't be a donkey. Keep your eyes and ears open and observe. Chickens do sneeze, cough, gasp, and rattle: but by the time you hear those noises, it's usually too late to do anything about it.

Any noticeable drop in the birds intake of water or feed consumption is a dead giveaway that something is wrong. It is up to you to try and correct the situation. If the condition is caused by extremes in temperature, very hot or cold weather, or a lack of proper ventilation, you can do something about it. If the chicks are listless and have no appetite, and this condition is caused by one of the diseases prevalent in chickens, once they are afflicted, there is not much that you can do. Therefore, prevention is the key!

INTERNAL PARASITES

Chickens are subject to many worms, but the one most likely to affect your flock is the large roundworm. This worm lives in the chickens' intestines and can cause the birds to lack vigor and their heads and shanks to become pale in color. The presence of the worm can cause a slump in egg production. The usual treatment is to use a deworming medication in the drinking water. Birds become hosts to the worm from picking at droppings or old chicken litter. As a precaution, don't use old chicken litter for bedding, and try to keep wild birds out of the chicken house.

EXTERNAL PARASITES

Lice and mites are the most common external parasites. A heavy infestation can cause loss of weight in the birds or a drop in egg production.

Chickens should be inspected periodically for evidence of these parasites. Ask your extension poultry specialist for advice in treating your flock for parasites. Always follow the manufacturer's directions carefully when using any dust, spray, or paint.

Lice. These parasites live their entire life on the chicken. They can be found on the skin under the wings, the breast, and under the vent. Thoroughly dusting the bird and the litter with a delousing agent is the usual treatment. Roosts can be painted with nicotine sulphate to kill lice.

Red Mites. These mites live on roosts and in laying nests during the daytime and come out at night to climb on the birds and suck their blood. They can be controlled by dusting the hens with malathion and/or painting roosts with mite paint.

Northern Fowl Mites. Like lice, these mites live their entire life on the chicken, and they are also found on the breast and vent. Dusting or spraying with the same chemicals used to eradicate lice will usually control them.

Many diseases can be prevented with a careful vaccination program. However, it is important that all the birds in your flock be vaccinated at the same time. If vaccinating by medication in water, make sure the birds are thirsty. Withdraw water for a few hours, and then give them the vaccine in a fountain or waterer.

Marek's Disease. Day-old chicks should be vaccinated for Marek's disease at the hatchery. Some hatcheries charge a small fee for the vaccination. It's worth the money. Marek's disease causes paralysis in birds. Because the birds don't eat or drink, they become emaciated. The mortality rate is high. There is no effective treatment for Marek's disease.

Coccidiosis. The symptoms of this disease are listlessness, ruffled feathers, head drawn into the shoulders, and acute, sometimes bloody diarrhea. The mortality rate is very high.

Coccidiostats should be provided in either the chick feed or water for the first 14 weeks of the chicks' lives. Coccidiosis can be treated with medication. Note: if you use sulfanilamides when treating laying hens, their eggs will not be fit for human consumption.

Infectious Bronchitis. The symptoms of this viral disease are coughing, sneezing, wheezing, and a nasal blockage or discharge. The hens have trouble breathing and stop laying. It can affect chickens of all ages and the mortality rate can go as high as 50 percent. There is no effective treatment and it spreads very rapidly through a flock.

To prevent infectious bronchitis, birds should be vaccinated when they are about 10 days old with medication added to their drinking water. The vaccine is often given in conjunction with the vaccine for Newcastle disease.

Newcastle Disease. This is another viral respiratory infection. The symptoms are coughing, sneezing, and paralysis. The hens stop laying. The mortality rate can be as high as 100 percent among affected birds. There is no known treatment.

Chickens can be vaccinated when they are 2 or 3 weeks old by medication provided in the drinking water. Pullets are usually vaccinated again about 1 month before they are due to start laying. In the case of lightweight breeds, this would be when they are about 16 weeks old. This gives them time to recover from the stress of vaccination before they go into production. Vaccination does cause stress in the birds, and their reaction time to vaccines can be from 5 to 7 days later.

If you keep layers over for a second year, they should be vaccinated again for Newcastle disease. Booster vaccinations of hens should be the same as the original medicine. Always follow the manufacturer's directions carefully. If you buy started pullets or older hens, ask about their previous vaccinations and proceed accordingly.

HANDLING DISEASE PROBLEMS

I do not recommend that you as a beginner, try to diagnose any disease in your flock by yourself. If in your small flock 3 or 4 birds appear sick at the same time, then it's time to panic and call your local extension poultry specialist for advice and direction. In my experience, these specialists have always been most helpful.

I suggest that you consult with your extension poultry specialist as to the most prevalent poultry diseases in your own immediate area before you embark on a vaccination program. It is not necessary to use a broad spectrum approach and vaccinate for all kinds of diseases that your birds may never encounter.

Only vaccinate or treat sound healthy birds. Don't try to medicate obviously sick birds.

If you follow all the rules with a small flock, disease should not be a problem. The USDA suggests that a commercial laying flock should have a minimum of 2,500 birds. There are many commercial flocks with 10,000 to 100,000 hens. As we all know, when you crowd huge populations of birds, animals, or people into constricted areas, you are inviting health problems. My own laying flocks never consist of over 25 birds, and I have never had any outbreak of disease.

MORE DISEASE PREVENTION TIPS

Don't keep chickens of different ages in the same pen. Older birds may be carriers of diseases and transmit them to the young birds. If poultry of various ages are kept on the same farm, their pens should be at least 50 feet apart.

Don't allow visitors, especially other poultry growers, to walk into your chicken pens. This may be very hard to do in the case of the children of neighbors, friends, and relatives who want to see the peep-peeps. Let them look through a window—or handle it as best as you can without too many hurt feelings.

Don't allow wild birds or family pets inside the chicken house. Screen the windows and keep all doors securely closed.

AT THE END OF THE HEN'S
PRODUCTIVE EGG-LAYING YEAR(S)

Sooner or later, the expense of feeding a certain hen can no longer be justified by the quantity of eggs she lays. The hen should be butchered and enjoyed as Sunday dinner.

Generally speaking, most hens reach the end of their productiveness after 12 to 14 months of egg laying. You can cull out the nonlayers a few at a time, or you can replace the entire flock (good layers and all) all at once to make room for a new young flock. Extra hens can be stored in the freezer, just as broiler-fryers are.

Dual-purpose medium-weight hens make good roasters. But even the lightweight hens can be put to good use in soup. Follow the same instructions for butchering and dressing egg-layers as broiler-fryers (pages 56–62).

Chapter 5

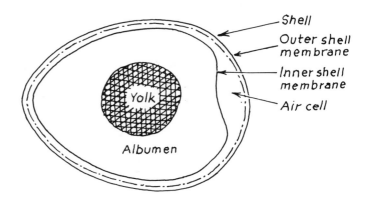

THE EGG

The hen's egg is an almost perfect food. It not only tastes good; it contains almost every food element necessary for human nutrition, and they are all in perfect balance. A human being could survive on a diet of eggs, provided he or she also ate citrus fruit. The only essential nutrient lacking in eggs is vitamin C. I am not, of course, recommending that anyone pursue a diet of this kind – but it is an interesting point.

The logo of the American Egg Board shows a hand holding an egg, with the words, "The incredible, edible egg." And the egg is incredible! It contains easily digested amino acids (protein), carbohydrates, fats, and important minerals such as phosphorus, potassium, iron, and magnesium. With the exception of ascorbic acid (vitamin C), the egg contains every other vitamin known to be needed by humans, including that most important vitamin B12, the lack of which can cause serious deficiency problems.

According to dietary standards set by the Food and Nutrition Board of the National Academy of Sciences, a serving of 2 large eggs at a meal supplies 30 percent of the daily protein and 6 percent of the daily calcium requirement for good human nutrition.

And, 1 large egg contains only about 80 calories. The larger the egg, the more calories it contains. Thus, a small egg will contain about 60 calories, while a jumbo egg will have about 100. The average egg contains 66 percent water, 13 percent protein, 11 percent minerals, and 10 percent fats.

The egg is one of the earliest of prepackaged foods, coming in its own individual container. And that container is a marvel of construction. The shape of the egg, with the shell fashioned in 2 convex arches, gives it great strength. It is almost impossible to crush an egg by squeezing its ends together with your hands. You can drop an egg from the rooftop of your house onto a lush lawn, and if it lands on end, it may not break. Egg dropping demonstrations are not infrequent. A popular television show, "That's Incredible," featured a segment in which eggs were dropped a distance of 300 feet from a hovering helicopter, and some of them survived the fall.

Who can disprove the ancient Chinese belief that an egg fell from the heavens and thus the world was created? The Phoenicians, Egyptians, and other venerable civilizations believed that the world was egg-shaped (oval) and was hatched from an egg made by the Creator.

An examination of the parts that make up the egg may give you a better understanding of how best to handle chickens for maximum egg production and how best to handle eggs for maximum quality.

THE SHELL

Although it has been said that it's very hard to shave an egg, the surface of the shell is not as smooth as it seems. The shell, composed mostly of calcium carbonate, is actually very porous. It contains thousands of minute pores, which enable the embryo within a fertile egg to utilize oxygen in the developing process and allows the baby chick to breathe shortly before it breaks out of the shell.

When first laid, the egg shell is enclosed in a protective covering called the *bloom*, or cuticle. This coating on the shell's surface helps to prevent bacteria from getting inside and also holds down the loss of moisture.

The size of the shell depends on the age of the bird. Young pullets lay smaller eggs than old hens.

The diet of the hens and their age dictates the strength of the shell. The smaller shell of young pullets' eggs is usually sturdier than that of older hens. Deficiencies in feed rations affect the quality of the shell. Hens that are free to roam the barnyard usually produce eggs with thicker shells than hens confined to cages or coops.

Contrary to popular belief, the color of the egg shell has nothing to do with its nutritive value or strength. The color of the shell depends on the particular region of the world that a breed or variety of chicken originated from. Thus, the Black Minorca, from the Mediterranean, lays a pure white egg. And the White Rock, an American variety, lays a big brown egg.

MEMBRANES

Two membranes, the inner shell membrane and outer shell membrane, are located just within the egg shell. As the warm newly laid egg begins to cool off, a small pocket, called the air cell, forms between the membranes at the larger end of the egg. On or about the fifteenth day of its development, the embryo turns its head towards the larger or blunt end of the egg, where the air cell is located. It is in keeping with the marvelous structure of the egg that the blunt end, where the chick's head is located at the end of its incubation, contains more pores in the shell than the small end. This greater concentration of pores in the shell makes it easier for the chick to breathe oxygen in and carbon dioxide out just before it hatches.

In the case of infertile storebought eggs, the air cell becomes larger and larger. Thus, the smaller the air cell within the egg, the fresher it is, and vice versa. This explains why a very fresh egg, being more solid, will sink to the bottom of a pan of water, while an older one may float.

THE EGG WHITE

The white is also called the *albumen*. More than half the protein of the egg is in the white portion. The older the egg, the thinner the white, which is 1 reason why an aging egg runs all over a frying pan.

When first laid, the egg white contains carbon dioxide. As the egg ages, the carbon dioxide escapes through the porous shell. If the white of the egg is cloudy, it indicates the presence of carbon dioxide, thus a fresh egg.

THE YOLK

This yellow portion of the egg contains all of the fats in the egg and something less than half of the protein. If a large egg contains about 80 calories, approximately 64 of those calories are in the yolk itself.

The color of the yolk does not affect its nutritive value. The color strictly depends on what the bird is fed. Hens that are fed rations containing yellow corn will lay eggs with yolks of medium to golden color. Hens that are on a ration of grains such as wheat or barley will lay eggs with lighter colored yolks.

CLASSIFYING EGGS

Eggs that are sold to consumers must be graded by size and quality according to standards prescribed by USDA. But even if you are not selling your surplus eggs, it is a good idea to know how the quality of your eggs matches up against the USDA standards.

Size

Eggs range in size from peewee to jumbo. They are classified by size according to the minimum weight of a dozen eggs. Table 5-1 lists the weights of these different classes.

Quality Grades

The quality standards classify eggs from AA to C. The gradings are based on the quality of the shell, the air cell, and the egg white. Table 5-2 lists the characteristics of the different quality eggs.

Notice that there is not much difference between a Grade AA egg and a Grade A egg.

A grade A egg must be packaged within 30 days of being laid. A 30-day-old egg may be fresh by USDA standards, but it can't compare with a fresh egg you gather from your own flock in the morning, then eat for breakfast. Once an egg is laid, it cannot get any better.

It is a tribute to our poultry industry, however, that they can supply us with billions of good clean eggs with amazing uniformity each year. Uniformity in eggs is taken for granted.

As Shakespeare wrote, "They say we are almost as like as eggs" (*A Winter's Tale*). Cervantes countered with, "As one egg is like another" (*Don Quixote*). But, if the truth be known, when you go out to the hen house each morning and collect the eggs, they will not necessarily be uniform at all. You may find broken eggs, cracked eggs, dirty eggs, clean eggs, green eggs, eggs with bumps and eggs with lumps, slippery eggs, and some with no shell at all (the latter condition probably caused by a calcium deficiency).

Table 5-1

US WEIGHT CLASSES FOR CONSUMER GRADES FOR SHELL EGGS	
Size or Weight Class	**Minimum Net Weight per Dozen (Ounces)**
Jumbo	30
Extra large	27
Large	24
Medium	21
Small	18
Peewee	15

Table 5-1. (From: Egg Grading Manual, *Agriculture Handbook No. 75, USDA, Beltsville, MD.*)

GATHERING EGGS

The sooner that eggs are gathered up after being laid, the cleaner they will be, and smaller is the chance they will be broken by accident. If possible, eggs should be gathered 3 times a day. If that isn't convenient, gather the eggs at least twice a day. If allowed to lie around too long, eggs can become soiled or cracked by a clumsy hen. Since it is the nature of a chicken to peck at least once at anything lying on the ground, some hens even try eggs and wind up liking them. An egg-eating hen should be culled.

Most important, prompt gathering of eggs aids in preventing rapid deterioration once they have been laid. The longer a newly laid egg stays warm, the quicker it loses its quality of freshness. Even the body heat of hens, should more than 1 bird use the same nest, can hasten deterioration.

Should you reach in under a hen on the nest in order to extract an egg or eggs? This is a moot point. Some hens won't pay any attention to you. Some will give you a sharp peck on the arm for your trouble. And others will jump up and run away to lay their egg another day. Large commercial egg factories, with their completely automated systems, don't have this problem. The caged hen lays an egg, and it drops down onto a conveyor belt and is carried away immediately. However, we are discussing the small flock here, and automation, even if desirable, is out of the question. Your alternative is to gather the eggs as frequently as possible.

HANDLING NONFERTILE EGGS

After you have gathered the eggs, clean any soiled ones and cool them all as rapidly as possible by putting the eggs in an egg storage room. Temperatures ranging from 50 to 60 degrees F. with a relative humidity of 75 percent are considered ideal for egg storage. Often an unheated pantry fits the bill just fine.

Dirty eggs can be immersed in a pail containing a detergent-sanitizer and then rinsed with clean water. The temperature of the cleaning solution should be from 110 to 120 degrees F. The eggs should not be left in the solution more than a couple of minutes, and then should be dried off right away. Detergent-sanitizer solvents are available from poultry supply houses and farm stores.

When there are only a few dirty eggs to deal with, soiled spots can be wiped off with a cloth dampened with a detergent-sanitizer and then rinsed with clean water. To be less technical, some small-flock owners wash dirty eggs with mild household detergents, then rinse them immediately.

Table 5-2

SUMMARY OF UNITED STATES STANDARDS FOR QUALITY OF INDIVIDUAL SHELL EGGS
(Specifications for Each Quality Factor)

Quality Factor	AA Quality	A Quality	B Quality	C Quality
Shell	Clean	Clean	Clean to very slightly stained	Unbroken
	Unbroken	Unbroken	Unbroken, may be slightly abnormal	May have slightly stained areas
	Practically normal	Practically normal		Moderate stains of less than ¼ of shell surface permitted.
				Prominent stains or adhering dirt not permitted.
Air cell	⅛ inch or less in depth	3/16 inch or less in depth	⅜ inch or less in depth	May be over ⅜ inch in depth
	May show unlimited movement, may be free or bubbly	May show unlimited movement, may be free or bubbly	May show unlimited movement, may be free or bubbly	May show unlimited movement, may be free and bubbly
White	Clear and firm	Clear and reasonably firm	Clear, may be slightly weak or thin	May be weak and watery
				Small blood spots or clots may be present
Yolk	Outline only slightly defined	Outline may be fairly well defined	Outline may be well defined	Outline may be plainly visible
	Practically free from defects	Practically free from defects	May be slightly enlarged and flattened	May appear dark, enlarged, and flattened before candling light
			May show definite but not serious defects	May show clearly visible germ development, but no blood due to such development; may show other defects which do not render the egg inedible
				Small blood clots or spots (aggregating not more than ⅛ inch in dia.) may be present

Table 5-2. (From: Egg Grading Manual, Agriculture Handbook No. 75, USDA, Beltsville, MD.)

The process of washing the egg does remove the natural protective coating, the bloom, and there is a decided loss of moisture from the egg through the shell and rapid deterioration of its quality of freshness. Unless washed eggs will be eaten within a short time, the shell should be coated with a colorless, odorless, tasteless, edible mineral oil that is available in aerosol spray cans.

Some small-flock poultry farmers don't wash their eggs at all, prefering to rinse them off just before eating them.

Obviously, any cracked or partially broken egg should be used immediately, preferably in baking.

EGG STORAGE

It is important to keep in mind that there are 3 main factors affecting the freshness of an egg.

- **Temperature: The newly laid egg should be cooled as soon as possible and kept at temperatures above 35 degrees F., but below 55 degrees F.**
- **Humidity: The egg should be stored in a cooler that maintains about 75 percent relative humidity. This helps in reducing the loss of moisture from the egg.**
- **Coating the shell: This aids in slowing down the further loss of moisture.**

How long can you keep an egg in storage and still have it retain any semblance of freshness? An unwashed egg probably keeps the best of all. If stored properly, it can keep up to 4 months. Nature's own protective covering, the bloom, works better in preserving the quality of freshness in an egg, than manmade products, such as mineral oils. A washed egg should keep for 4 or 5 weeks in a refrigerator, without losing too much quality.

When storing eggs in a refrigerator, keep them in a carton, large end up, and place them in the coldest area of the appliance. Do not store them next to the onions or other vegetables with pungent odors. Eggs absorb odors easily, and it will affect their taste.

FREEZING EGGS

When your chickens are producing more eggs than you can eat fresh, you can freeze the surplus. Freeze only clean fresh eggs. Don't wait until your refrigerator is overflowing with aging eggs before you get around to freezing them. The fresher the egg that is frozen, the higher its quality.

Whole Eggs and Yolks

When yolks are frozen, they thicken, or gel. To help retard this gelation, you can add either 1/8 teaspoon salt or 1½ teaspoons sugar to every 1/4 cup yolks (about 4 yolks) or every 1/4 cup whole eggs (about 2 eggs). Freeze in plastic freezer containers and be sure to label the container with the number of eggs or yolks, and whether you have added sugar or salt. This way you will know how many eggs you are defrosting and whether you will have to adjust your recipes in any way.

It is a good idea to freeze eggs in small quantities. Two eggs are often required for baked goods, such as cakes. Freeze some eggs singly, if you have the space. Frozen whole eggs are best used in baking.

Before using a frozen whole egg or yolk in a recipe, thaw in the refrigerator or under cold running water.

Whites

To freeze egg whites, break and separate the eggs. Be sure that no yolk gets in with the whites. Whites can be frozen without any added ingredients. Pour them into freezer containers, seal tightly, label with the number of whites included, and freeze. Thaw the frozen whites under refrigeration or cold running water. You can use the thawed whites just as you would fresh egg whites.

SOME FINAL THOUGHTS

Years ago, when the majority of our population was more rural that urban, most every family kept a small flock of chickens for the eggs they produced. Since tending chickens and gathering eggs was a normal part of daily life, our figures of speech were very much influenced by references to hens and eggs. And the language was certainly the more colorful for it. Some of the sayings are still very much in use today. We still egg people on, crack ideas in the egg, and walk on eggs. Presidents appoint eggheads to administrative jobs. We encounter bad eggs and good eggs in our personal relationships. In this age of the Individual Retirement Accounts, we have the nest egg – not to forget the comedian who bombs out and winds up with egg on his face.

In regard to which came first, the chicken or the egg, perhaps Samuel Butler said it all: "It has, I believe, been often remarked, that a hen is only an eggs' way of making another egg."

Part Two: Ducks

Chapter 6

WHY
RAISE
DUCKS?

The best reasons that I can think of for raising ducks are that they are fun, colorful, interesting, much hardier than chickens, and are good providers of meat and eggs.

With proper management, a duck raised for meat can attain a live weight of 7 pounds in just 7 weeks. In contrast, it takes about 8 weeks to raise a broiler-fryer chicken to a live weight of 4 pounds. Some strains of egg-laying ducks average over 300 eggs per year (under excellent management), a higher rate than most breeds of laying chickens.

Duck has always been a popular source of food in Europe and Asia. But in this country, the increase in total duck production has only shown a modest gain from year to year. In my personal opinion, it's our loss.

DUCK MEAT AND EGGS

Duck meat is delicious! And it doesn't have to be greasy, as it is commonly held. The meat *can* be fattier than that of chickens or turkeys, but this depends on the breed of duck, the diet the duck was raised on, and the way the bird is prepared. You can remove any fatty deposits and even skin the carcass before roasting it to eliminate excess fat. Ducks that are raised partly on forage, as I recommend, will be much leaner than ducks raised in total confinement, as most commercially produced ducks are.

In contrast to the chicken, which provides us with both white and dark meat, the carcass of the duck contains all dark meat. This is due to the presence of myoglobin, a pigment in the duck's muscles that stores oxygen and helps sustain them when they fly long distances. With the exception of the female Muscovy, our domestic ducks are poor flyers, but this carryover from their wild ancestors still colors their meat. When we eat meat, whether poultry or 4-legged animals, we are eating muscle.

If you have never eaten duck, it would be best if you tried it to see if you liked it before you get into raising ducks. Duckling, whether fresh or frozen, is readily available at most grocery stores. Buy a duck, take it home, and make the taste test. If, for some reason, you don't find it delicious, don't go any further, unless you want to raise ducks for exhibitions or as a hobby.

It would be wise, if you are contemplating raising ducks for eggs, to make the same taste test. Duck eggs are not so readily available. I have never seen them offered in any store. The best way to locate a source for duck eggs is to ask at your closest farm feed store for the names and addresses of duck growers in your neighborhood. Your extension poultry specialist also can help in locating duck growers. Then buy a dozen duck eggs and try them out in various ways: fried, soft-cooked, scrambled, in omelet, and in baking. You will probably find that duck egg whites are much firmer than chicken egg whites.

By the way, there is no commandment carved in stone that says, "Thou must have orange sauce with duck!" Many fruits, including apples and pineapples, accent the delicious flavor of roasted duck meat. Ducklings also can be fried, barbecued, fricasseed, stewed, stuffed, creamed, and served sweet and sour. And there's always duck soup.

Having eaten and enjoyed duck meat most all of my life, I was quite surprised to find out, upon conducting a small survey, that only 1 out of the 10 homemakers I queried had ever eaten duckling. One lady had a duck in her freezer and another said that she'd thought about it a few times. The lady with the duck in her freezer had kept it there for several months, not being sure as to how to prepare it. Her husband had suggested roasting it in a paper bag. It's probably a good thing that the bird was still in the freezer because I don't know how the paper bag would add to the flavor except to smolder or burn. I suspect his ulterior motive was to keep the interior of the oven free from spattering grease. A better idea would be to have the duck on a rack and put a small amount of water in the roasting pan below.

There's nothing complicated about cooking duck. Allow about 20 minutes per pound for an unstuffed bird; and roast the duck on a rack in a roasting pan at 350 degrees F.

DOWN AND FEATHERS

In addition to producing meat and eggs, ducks can provide down and feathers that are most valuable in making clothing, such as vests and jackets, and bedding.

INSECT CONTROL

Ducks are very good for insect control in lawns and gardens as their long, broad bills are more adept and better fashioned for catching flying insects, such as grasshoppers, than the narrow, hooked beak of the chicken. Vegetable garden experts tell us that one way to control slugs in your garden is to bait them with small saucers of beer. A more productive way to get rid of slugs is to let ducks eat them and gain valuable protein at the same time.

SHOW DUCKS

Ducks also are raised for showing at poultry exhibitions, as a hobby, and sometimes just to provide a conversation piece—they look so handsome when they swim effortlessly on a small pond.

CONTROL WEEDS IN PONDS

In case you wondered, it is not necessary to have a pool or pond for raising ducks. They can do quite well when raised on land, provided they have access to plenty of drinking water at all times. Ducks are good foragers, much better than chickens, although not nearly as good as their long-necked cousin, the goose. When placed on range, their intake of green feed, weeds, seeds, and bugs can supply them with 10 to 20 percent of their food requirements. If kept on or near small bodies of water, ducks will clean ponds of algae and other undesirable growth.

WHAT'S INVOLVED IN RAISING DUCKS

Raising ducks for meat is not a long-term commitment. Even as a beginner, you will probably be able to get them up to good weight within 8 weeks or so. If you don't enjoy working with ducks, you can get out of the business in a couple of months.

Raising ducks for their eggs is a longer process. Starting from scratch, it will take 5 to 7 months before the hens lay their first egg. If you hang in there that long and survive, the experience will probably make you a dyed-in-the-wool, or feather, duck person forever.

Raising ducks is a 7 day a week job. You can't leave domestic ducks to fend for themselves. You must feed and water them, see to their physical comfort, and protect young birds from inclement weather and predators. However, the time required in caring for a small flock of ducks is only about 15 minutes, twice a day, morning and evening.

Also, you don't need elaborate buildings or a lot of land in order to raise ducks. In fact, if your goal is producing the most meat in the quickest possible time, ducks are best kept in confinement or semi-confinement. However, I recommend that they be given access to a yard or pasture, so that they can benefit from sunshine, fresh air, and a varied diet, whether they be raised for meat or eggs.

You will need some kind of building for your ducks. A barn isn't necessary. A simple shed or garage is all that is required. Equipment is minimal: feeders, waterers, and supplemental light and nest boxes for egg-laying ducks. Also the duck yard should be fenced.

An advantage of ducks over chickens, especially if you live in a suburban or semirural area, is that a drake (male duck) speaks in a low, throaty voice; and unlike a crowing rooster, doesn't wake the neighbors at 5 A.M. (The one exception to this is the Muscovy drake, which hisses rather than quacks.) Some female ducks, particularly those of the White Pekin breed, are noisy, but no more so than nervous Leghorn chickens. Although, for some reason, a quack seems to carry further than a cluck.

If you are considering raising ducks for meat, it is important that you or members of your family attend to the butchering and dressing out of the ducks when they reach good slaughter weight. If you feel that you are not capable of killing and cleaning the birds yourself, then perhaps you should raise ducks as a hobby, or go with the egg-laying varieties.

If you are going to eat the birds, don't make pets out of them. You can't kill and eat a pet. And, if you raise ducks for meat, you will need a freezer with large storge capacity, at least 75 pounds.

SEVEN FACTS ABOUT DUCKS

- Ducks are very hardy and less subject to disease than chickens, if cared for properly.
- The average productive life of a duck kept as an egg layer or breeder is 2 to 3 years. In comparison, the average productive life of a chicken is 1 to 2 years.
- Being waterfowl, ducks can survive and prosper in damp or rainy climates.
- If given ample shade and drinking water, ducks can do well in hot weather.
- **Ducks can take extremely cold temperatures.** Sometimes, when out driving in late fall or early winter and basking in the comfort of our well-heated car, we shiver at the sight of ducks out on a lake and wonder how they can survive the icy water. Well, there is an explanation. Their under-feathers are a very heavy down, close to the skin. Above these are an overcoating of contour feathers, so-called because they form the general covering of the duck and determine its external shape. The contour feathers trap air underneath and between the down and the skin. This serves as insulation and helps keep the bird warm. Also, ducks are covered with down and feathers from head to legs and have no extraneous appendages, such as combs and wattles, that will freeze.
- **Ducks are built to live on the water.** Their undercarriage is fairly flat and wide and designed for floating. Their feet are webbed, providing a powerful thrust for swimming. They have an oil gland on their back, adjacent to the base of their tail. When preening, they turn their head and neck about 180 degrees and squeeze this gland with their bills, spreading oil about their feathers and making them waterproof.
- **The legs of the duck, with the exception of the Indian Runner breed, are fairly well amid-ship and placed far to each side of the body.** Thus, when walking, they alternately place each foot inward and toward their center of gravity, which produces the characteristic waddling motion.

These legs, which make them look clumsy when walking on land and serve them so well when they swim, are rather delicate, particularly at the joint where shank meets thigh, and this is an important point. You can grab chickens by the legs and pick them up and even buy a long-handled crook for catching them by the legs. It is not, however, a good idea to grab ducks by the legs and pick them up that way. When catching ducks, it is best to grasp their neck with one hand and cover or hold down their wings with the other hand. Once caught, they can be held with one hand underneath their body, supporting its weight and holding the legs, their head tucked under the arm, and the bird held gently but securely against the waist.

RAISING DUCKS WON'T SAVE YOU MONEY

Do not, repeat, do not, decide to raise ducks for meat with the notion that you will save money. You probably won't. Let's look at some figures.

A baby duckling will cost at least $1.50, depending on the variety. The postage cost of 1 duckling sent from a distant hatchery will be about $.35. That duckling will eat about $2.50 worth of feed in order to gain 7 pounds. Adding that up, the total comes to $4.35.

I can buy a well-prepared, frozen, 5-pound duck at my local supermarket for $1.29 per pound, or $6.45. The frozen duck has been butchered, eviscerated, drawn, the feathers and down removed, any stray hairs singed off, the giblets prepared and washed, and the whole thing is attractively packaged. Subtracting $4.35 from $6.45 leaves a balance of $2.10. If you add up the cost of building maintenance, electricity for lighting, a feeder, a water fountain, bedding (if purchased), and the time and effort involved in carrying water to the duck pen from your house or another source, plus the half hour each day necessary for caring for the ducks, it has to come to more than $2.10.

Baby ducklings cost more than baby chicks initially, and they eat more feed during their growing process.

You won't save money raising a small flock, but you will gain in satisfaction. Knowing that by your hand your family was fed is a good feeling.

Chapter 7

SELECTING
A DUCK
BREED

Once you've decided to go ahead and start a home duck flock, your first decision is which breed to raise.

As with chickens, you must decide whether to raise ducks primarily for the meat or for the eggs. If you like the meat of the duck, but not the eggs, the decision is easy: raise meat-type ducks.

If you decide to raise an egg-laying type, you will still end up with duck meat. Since I advise buying day-old unsexed ducklings, half your birds will be males. The drakes should be grown to about 7 pounds in 8 weeks or so, then butchered. The hens will start laying eggs at 4 or 5 months of age.

If you do decide to raise ducks for eggs, keep in mind that hens can average about 300 eggs a year. It is advisable to keep only enough ducks to supply your own family with eggs, unless you have a known market for your excess eggs or you plan to hatch out your own ducklings. In general, there is very little demand for duck eggs in the United States.

White Pekin ducks are favored by the commercial duck industry for their white feathers, yellow skin, and ability to gain weight rapidly. (Photo courtesy USDA)

White Runners are excellent egg-layers as well as good foragers and graceful runners.

Khaki Campbells can average over 300 eggs per year, exceeding the output of most of the best laying breeds of chickens. (Photo courtesy USDA)

The White Pekin

Originating in China, the White Pekin was brought to the United States in the 1800s. It is by far the most popular and populous meat duck. The majority of the ducks raised in the United States are White Pekins, and about 60 percent of these are grown on Long Island, New York. Hence, when you buy a frozen "Long Island duckling" at your supermarket, or order one at a restaurant, you will probably dine on a White Pekin.

The White Pekin is favored by the commercial duck industry for many reasons. It is the fastest gaining of all ducks and can reach a live weight of 7 pounds in just 7 weeks under expert management. It has white feathers. When dressed out, the bird presents an attractive carcass, due to the absence of the tiny, colored pinfeathers which may leave dark splotches in the skin when plucked. The bird has a yellow skin, which is desired by the American consumer. Of all the heavy meat types, the White Pekin is the best egg layer, with hens averaging 120 to 160 white-shelled eggs per year.

The temperament of White Pekins tends to be somewhat nervous, and the ability to set eggs or brood a flock of ducklings has been bred out of them. White Pekins are not particularly good foragers and are best kept in semiconfinement to accomplish their main purpose—producing the most meat in the shortest time.

Mature adult Pekin drakes will weigh about 9 pounds and the hens about 8 pounds. They are readily available as day-old ducklings.

White Pekins tend to be fattier in carcass than most other ducks, but not all of this can be blamed on the breed. White Pekins are often kept in confinement. They are often fed a high-energy ration in the latter stage of their growth in order to get them up to good slaughter weight, but this diet causes them to put the fat on. Ducks that forage for part of their feed have leaner meat.

Aylesbury

The large, white-feathered Aylesbury originated in England. It is about as fast growing a breed as the White Pekin. With proper management, the Aylesbury can achieve a weight of 7 pounds in 8 weeks. Having white plumage, Aylesburys present an attractive carcass when dressed out. Their skin, however, is white, and the American consumer appears to prefer a bird with yellow skin. Adult drakes will weigh about 9 pounds and hens about 8 pounds.

Aylesburys produce fewer eggs than White Pekins, with the hens averaging 50 to 100 white or tinted eggs annually. They are not particularly good foragers and show a lack of interest in setting eggs or brooding baby ducklings. Also this breed is not readily available. You have to search far and wide to find a hatchery that handles them.

Rouen

In my opinion, the Rouen is one of the most handsome of all ducks. Originating in France, the Rouen has the same color patterns as the wild Mallard but Rouens have much more vivid and brighter hues. (The Mallard is considered the ancestor of all domestic breeds of ducks.)

An adult Rouen drake has a green head with a white collar around its neck. The breast feathers are a claret color, and the underbody is grayish. There is a patch of blue on the wings. The female Rouen duck is more subdued in color, except for a blue patch on her wings. Her feathers are a rich brown, with black penciling on the breast. The Rouen has a yellow skin.

Rouens are excellent foragers, better than White Pekins. They are easily available. An adult drake will weigh about 9 pounds and the hen about 8 pounds.

Rouens produce fewer eggs than Pekins. The hens average 50 to 125 white or bluish-green tinted eggs. When dressed out, the carcass may be less attractive due to the dark colored pinfeathers, which can leave marks in the skin when plucked. This is one reason they are not favored by commercial duck growers. However, these birds are perfectly fine for home consumption. They do not gain as fast as White Pekins. One reason for this is that they are not usually *pushed* as fast as the Pekin and are allowed to forage for some of their food, which in turn makes for a leaner bird. They can reach a good slaughter weight in 10 to 12 weeks.

Muscovy

The Muscovy, which originated in South America, is not considered a true duck for many reasons. They don't act like most other ducks and seem to march to a different drummer. They fly more than other ducks, especially the females. They have a tendency to roost. Their feathers are not as waterproof as those of other ducks; if kept too long in the water, they can drown. Also, there is a marked sex difference in the weight of the adult birds, with drakes weighing up to 12 pounds and hens weighing about 7 pounds.

Not the prettiest of ducks, Muscovies have patches of coarse red skin around their eyes, and the males have red caruncles (fleshy growths) on their faces. They come in many varieties and colors, including white (the variety most popular with growers aiming for the commercial market), colored (black and greenish-black overall, with white patches on the wings), chocolate, buff, silver, and pied (mottled black and white).

They are excellent foragers and have strong maternal instincts in regard to setting eggs and brooding large numbers of ducklings. Also, they aren't likely to disturb the neighbors; the females are almost mute and the drakes hiss rather than quack. The meat of the Muscovy is leaner than that of most other ducks.

Muscovies have been crossbred with White Pekin ducks in China for several centuries. This cross has led to a variety that has the fast-growing traits of the Pekin and the leaner meat of the Muscovy. Muscovies have been crossed with Mallards with favorable results. The hybrid offspring of these crosses are sterile and called mules.

When dressed out, the carcass has a pinkish-white skin. Muscovies are readily available for purchase in the United States.

The disadvantages of this breed is that the hens' output of eggs is meager; the hens average 45 to 100 white or tinted eggs per year. They have a tendency to fly, particularly the lighter females. Muscovy eggs take about 35 days to hatch out, in contrast with an average of 28 days for the eggs of other duck breeds to hatch. They are slower growing than most other breeds. However, this can be a plus in that they are less fatty.

If fed a light ration of grain and allowed access to good pasture or range, Muscovy drakes can gain up to 10 pounds in about 4½ months and the females weigh about 5 pounds within that time period. The slow rate of gain, plus the big difference in size and weight between males and females, makes them unsuitable for mass production as meat ducks.

THE EGG BREEDS

Khaki Campbell

Khaki Campbell is a high-producing breed that originated in England as a cross of Indian Runner ducks, Mallards, and Rouens. The drakes have greenish-bronze heads; their necks, backs, and tails are brownish-bronze; and the rest of the body feathers are of khaki color. Their bills are green and their legs and feet are orange. The hens have brown heads and necks, while the rest of their coloring is khaki. Their bills are a greenish-black and their legs and feet are brown. There is also a white variety of Campbell duck, which is not as popular as the original, darker khaki-colored duck.

Some strains of the breed will average well over 300 nice white eggs per year, exceeding the output of most of the best laying breeds of chickens. I like duck eggs; however, in my personal opinion, the smaller the egg, the better its taste. A Khaki Campbell egg will average about 2½ ounces in weight, which is small, just slightly larger than the egg of a Leghorn chicken. The eggs are fine for frying, steaming, poaching, scrambling, and making into omelets. A Muscovy egg will weigh about 4 ounces and a White Pekin egg will average about 3½ ounces. These larger eggs are probably more suited for baking than for table fare, although I enjoy them both ways.

Khaki Campbells are excellent foragers when put on range. They are fairly easily available from hatcheries within the country.

As a rule, Khaki Campbells are not good for setting eggs or brooding baby ducklings. Due to their relatively small mature size—adult drakes and hens average about 4½ pounds—they are not usually raised for meat. However, if well fed, they can gain 3½ to 4 pounds in about 8 weeks and make good, if small, roasters. An added bonus is that the meat is fairly lean.

Indian Runner

The tall, slender, erect-standing Indian Runner originated in the East Indies. Their egg-laying capacity was highly developed in Europe, and they are second in egg production only to the Khaki Campbell. The hens of some strains of Indian Runners will lay 200 to 250 white or tinted eggs per year.

The most common varieties of this breed are the White, Penciled, and Fawn and White. Due to their upright carriage, a very long neck, and feet placed well aft, they don't waddle like other ducks, but run quite gracefully.

They are excellent foragers and their availability from breeders and hatcheries is good.

Despite their prodigious egg-laying capacities, Indian Runner hens do not have strong maternal instincts and they are not good prospects for either setting eggs or brooding baby ducklings. Mature drakes and hens will weigh only about 4½ pounds. Obviously they are not suitable to be raised for heavy meat production. The males can be utilized as light roasters, dressing out to 3 to 3½ pounds, with the hens being kept for laying purposes.

Chapter 8

RAISING DUCKS FOR MEAT

Having decided that you want to raise ducks for meat, your next step is to make a final decision on a breed and order your stock. Then you must prepare for the arrival of your ducklings.

START WITH DAY-OLD DUCKLINGS

It is possible to purchase fertile duck eggs suitable for incubation from hatcheries, but they usually cost about $1 apiece, plus transportation charges, and there is no guarantee for their safe arrival. Also, the hatching percentage for duck eggs, particularly for beginners, averages from 50 to 60 percent, so you will have to purchase 25 to 30 eggs to ensure that you have 15 ducklings. Then you have to buy a mechanical incubator, or else purchase broody hens, either chickens or ducks, and wait out the egg hatching period of some 28 days.

You can also purchase adult ducks from some hatcheries. A trio of White Pekins will go for about $37.50 and a pair of Rouens will cost about $65. You then have to wait until they mate and hope the resulting eggs are fertile, and hope the females are inclined to set the eggs for 28 days, and then hope that they brood their hatch until the ducklings are well-feathered out and don't need a mother's warmth anymore. Also, White Pekin hens usually are not inclined to set eggs or brood ducklings. The Rouens are pretty good at both jobs.

In any case, this seems like a long way to go about it. Besides, half the fun of raising ducks is watching them grow up from fuzzy little balls of down into handsome birds. There are no ugly ducklings.

It makes more sense to buy day-old ducklings of the most common meat breeds. They will cost about $1.50 apiece, plus postal charges, and are guaranteed 100 percent live arrival, and you can't beat that!

THE BEST TIME TO BUY DAY-OLD DUCKLINGS

It is a good idea to purchase day-old ducklings in the spring or early summer. The temperate weather of spring and early summer can only help when you are raising ducklings for the first time. In any case, except for White Pekins, most breeds of ducks are only offered for sale from March through July or August.

I suggest that you place your order with a hatchery at least 1 month ahead of time and request a June first delivery.

WHERE TO BUY

Your local extension poultry specialist is probably one of the best sources of information about reputable hatcheries. It is extremely important that you buy the best ducklings from the best stock available. Bargain birds are usually no bargain. (See the appendix for names and addresses of extension poultry specialists, listed state by state.)

Rural newspapers and farm magazines are another good source of suppliers of waterfowl. (See the appendix for poultry publications.) The National Poultry Improvement Plan, a service provided by the USDA, in cooperation with state authorities, will provide you with a directory of participants who handle waterfowl. (See the appendix.)

If possible, buy your ducks locally. Besides saving on delivery charges, you are close to the source if problems arise.

WHAT KIND OF DUCKS TO BUY

I suggest that you choose from either the White Pekin or Rouen breeds. The Pekins are most readily available and will achieve a good slaughter weight quickly, even when suffering the mistakes that most novices make. The Rouens take a little longer to reach optimum weight, but are better foragers. If given access to a pasture or range, Rouens will provide leaner meat.

HOW MANY TO BUY

The minimum number of ducklings to buy is no less than 15 unsexed day-old ducklings. Unsexed means that they are sold as-hatched, both males and females. The females grow a little slower than the males, but taste just as good, and it is almost impossible to purchase sexed ducklings. If a hatchery takes the time and trouble to determine the sex of newborn ducklings, they will charge you at least $.50 more per bird. With a small flock, the extra cost just isn't worth it. A flock of 15 ducklings butchered at a weight of about 7 pounds each will provide you with about 75 pounds of meat, bone included.

APPROPRIATE HOUSING

Before the day-old ducklings arrive, you must see to providing adequate housing. Their shelter needs are not extensive: just a warm, dry, draft-free place for the first few weeks of their lives. You can house the ducklings in any kind of building that will provide these requirements, whether it's a barn, shed, old poultry house, or garage. With present day construction costs, it isn't reasonable for a beginner to erect a brand new building for a small flock of ducks.

When they are small, the ducklings won't take up much space, but by the time they are 6 to 8 weeks old, they will need a minimum of 2½ square feet of indoor space each. It is wise to provide for their ultimate space requirements at the start. Thus, 15 ducklings will eventually need about 37½ square feet, or a space about 8 feet by 5 feet in a house or pen. Twenty ducks will need about 50 square feet of space, and so on. You will also need room for a feeder and waterer and enough space for you to walk around in without stepping on the ducklings.

While I firmly believe that once ducklings are 4 weeks old, they should have access to an outside yard with all of the benefits a range provides, I also believe that they should be shut in at night. The main reason to confine meat ducks at night is to protect them from predators. With egg-laying ducks there is a second reason for confinement: to be sure they lay their eggs where you can find them.

If you are going to use a barn to house your ducklings, but it's very large, you can partition a section of it off with plywood or chipboard. There is no reason why you can't use large buildings. In the beginning, however, you will confine the young birds to a very small space with the use of a protective guard. If you do partition off a barn or large building for the ducklings, it would be best to utilize the southernmost side to take advantage of the most temperate natural exposure.

The building can have a dirt floor, but wooden or cement floors are much more desirable because they can be cleaned. No matter what type of flooring, in the beginning you will provide several inches of litter or bedding to keep the ducklings warm and dry.

All doors and windows must close tight and secure. A fine-meshed wire should be nailed on the inside of the windows to keep wild birds from flying inside the house and possibly introducing diseases to the ducklings.

You will need electricity in the building. If the building isn't wired, you can run a heavy-duty extension cord from the nearest source of power to carry heat and light to the birds.

In the beginning, fortunately, the same pen or house can be used for the ducklings, whether they will ultimately be slaughtered for meat or raised for egg production.

When the ducklings are about 4 weeks old and are substantially feathered out, you will want to turn them out to a yard where they can benefit from fresh air, sunshine, weeds, seeds, and insects. At this stage you will still have to keep a watchful eye on them and chase them back into the house at the first sign of a storm, high winds, or cold rain.

The duck yard or pasture should have good drainage, as pools of stagnant water are breeding grounds for disease. Provide shade for very hot days and plenty of drinking water at all times.

The fencing for the pasture doesn't have to be very high; most domestic ducks are not good fliers. Two-foot-high woven or sectional wire will do the job of keeping the ducks in, but it won't keep large predators, such as dogs, out. A fence 5 feet high, with a strand of barbed wire at 2 inches above ground level on the outside of the fence posts, and a strand of barbed wire at the top of the fence, will keep large predators out.

The size of the yard is arbitrary. It all depends on the terrain and ground cover. It is recommended that mature ducks be provided with at least 10 square feet each on pasture. Your young ducklings should have at least half that area. If there is an abundance of short green succulent vegetation, they will need less space.

In my opinion, a duck yard or range cannot be too large. But even a small yard is better than no yard at all.

NECESSARY EQUIPMENT FOR THE DUCK HOUSE

The equipment needed to brood day-old ducklings is just about the same as is required for brooding chicks. You will need a 250-watt heat lamp to provide warmth for the ducklings for the first few weeks. You will also need a feeder, a waterer, litter for bedding on the floor of the house, and a small guard or corral to keep them close to the source of heat, food, and water.

The 250-watt heat lamp should have a safe porcelain socket, a reflector above the bulb and a metal band below. You will also need a thermometer to monitor the air temperature near the ducklings.

To start out, ducklings will need about 1 inch of space at the feed trough. The feeder can be made of metal or wood. It should have a roller or spinner bar running lengthwise across the top, so the ducklings cannot get into the trough and soil the feed. The edge of the trough should be at about the same height as the little birds' backs so they can reach the feed comfortably. A feed trough 12 inches long and open on both sides will supply plenty of room for 15 ducklings for the first few weeks of their lives. Chicken feeders work fine. (See page 43.)

A 1-gallon waterer will provide more than enough water for 15 ducklings for the first month. The waterer should be deep enough so that the ducklings can submerge their entire bill into the water, but constructed so that they cannot climb inside and possibly drown.

You will need a supply of litter to serve as bedding for the duck-lings. It helps keep them warm and dry and absorbs droppings. Chopped straw, coarse sawdust, wood shavings, or peat moss can be used for litter. Whatever material is readily available and economically priced is best.

Finally, you will need a guard or corral. It can be made of corrugated cardboard and should stand at least 1 foot high. It will be placed under the heat lamp in a circle. (See page 43.)

PREPARING THE DUCK PEN FOR THE ARRIVAL OF THE DUCKLINGS

Whatever type of shelter you use for housing the ducklings, it must be as clean as possible. Unless the building is new, this means that you have to brush and sweep the ceiling and walls and scrub the floor. If you are going to use an old poultry house, buy a disinfectant and use it on the walls and floor. Follow the directions carefully and air the building thoroughly so that no chemical residues are left to harm the ducklings. Poultry disinfectants are available at most feed and farm stores.

When the building is clean, hang the 250-watt lamp in the center of the pen, using wire, chain, or rope attached to rafters or the ceiling. The lamp should hang about 20 inches above the floor. The heat lamp will have to be either raised or lowered from time to time to adjust the temperature, so the suspension system must be easily maneuvered up or down.

A day or so before the ducklings are to arrive, test out your brooding arrangement. Put litter on the floor to a depth of 3 or 4 inches, turn the 250-watt heat lamp on, and place the guard or corral in a circle underneath the lamp and about 3 or 4 feet away from the center of the heat rays.

For the first week of their lives, the day-old ducklings will need a temperature of about 90 degrees F. in their pen. In order to estimate the temperature, place the thermometer at ground level within the corral guard. After a period of 3 or 4 hours, check the reading on the thermometer. If the temperature is above 90 degrees F., raise the heat lamp. If the temperature is below 90 degrees F., lower the lamp. Each time you move the lamp, wait at least 2 hours before making another change. When the ducklings arrive, they will be your best indicator as to whether your setup is correct, too hot, or too cold. In the meantime, however, you are ready for them.

Place the feeder and waterer adjacent to, but not directly under, the heat lamp and within the confines of the corral. Be sure to have feed on hand, but don't fill the trough or waterer before the ducklings arrive.

Ducks are messy eaters and use a lot of water when they eat. When eating dry feed, which you will feed them, they tend to slop a lot of water around in the process of swallowing and cleaning their bills and nostrils. This, of course, causes the litter around the water fountain to become damp. Damp litter invites mold and causes disease. To counteract this, you can place extra litter around the waterer and plan to remove the wet bedding periodically. Another method is to have a drain underneath the water fountain or else place the fountain on a wire screen with lots of litter below.

I have always enjoyed watching animals and birds eat their food with obvious relish and gusto. In the capacity to put the groceries away, I had always considered the pig to be the champ — until I saw a hungry White Pekin duck shovel it in. (There is a wild duck named the Shoveler. This duck is probably termed Shoveler more because of its broad, spoon-shaped bill, rather than the speed of its feeding habits.) Anyhow, once a duck positions itself and draws a bead on the feed, the rapid movement of neck, head, and bill in devouring food is like an automatic machine. Relative to size and weight, in an eating contest with a pig, a duck would win hands down.

WHAT TO DO WHEN YOUR DUCKLINGS ARRIVE

If your day-old ducklings are coming from a distant hatchery, they will be shipped via Air Parcel Post. Your local post office will notify you when the ducklings have arrived. Open the shipping box in the presence of a postal employee and count the living birds. All reputable hatcheries insure their shipments and will offer a refund or make good on your original order. If any of the ducklings have died enroute, file a claim with your postmaster and send it to the hatchery.

When you bring your ducklings home and into their own house, remove them from the crate one at a time, and dip their bills into their water fountain, which should be filled with lukewarm water. Then dip the bill of each duckling into the feed. The trough should be half full with duck starter ration. If the ducklings have trouble eating out of the trough this first day, you can sprinkle more feed on small shallow box tops or jar lids scattered about their pen. Within a few days the ducklings will get the hang of eating out of the trough, and you can remove the box tops and jar lids.

BROODING THE DUCKLINGS

For the first week of their lives, the day-old ducklings require a temperature of about 90 degrees F. Even if you have been able to maintain that temperature in their pen by raising or lowering the heat lamp, the best indicator of the ducklings' comfort is their actions. If they are all crowded under the lamp, the ducklings are cold. If they are crowding the sides of the corral, they are too hot. If the ducklings are all huddled against one side of the guard, there is a big draft, which must be eliminated. If they are spread around the pen, eating, drinking, and darting about in duck fashion, the temperature is just right.

Each week, the temperature should be lowered by about 5 degrees. You will do this by raising the heat lamp, always observing the actions of the ducklings and being aware of their comfort. The second week, then, the temperature should be at about 85 degrees F. The third week it should be about 80 degrees F. And so on. By the fourth week, the temperature should be at 75 degrees F., and the ducklings should be well feathered out. You can begin to turn them out in their yard or range at this time on warm, sunny days.

By the fifth week, the ducklings should be allowed to go outside at will, provided they are given shelter in inclement weather. At this stage, they should have sufficient plumage to survive the normal climactic conditions in most of North America in early summer and not require artificial heat from the lamp.

When you confine the birds within their house at night for reasons of security, it is a good idea to keep a low wattage bulb turned on all night. This will aid the birds in finding water or food during the night and help prevent them from stampeding and injuring themselves which ducklings are prone to do if they hear loud noises or thunder.

OBSERVATION AND DAILY CARE

Keep feed and water available to your ducks at all times. If they don't clean up all the feed, remove stale feed before it becomes moldy and replace it with fresh feed. Rinse the waterers out at least once a day. Due to the nature of their feeding habits, ducklings can muddy up a water fountain minutes after you have added fresh water. You can't keep the waterer sparkling clear all the time, but frequent replenishment will help.

With a small shovel or dustpan, remove droppings and soiled litter from the pen every day. Add fresh bedding when needed. Pay particular attention to the area around the water fountain and keep it as dry as possible.

Remove any obviously sick ducklings from the flock and incinerate or dispose of them by burial, at least 1½ feet underground.

Observe your birds every time you feed and water them. You will soon develop a practiced eye and a nose for trouble. Don't call a vet for a single sickly duckling; but if a sizable portion of your flock becomes listless, droopy, doesn't eat or drink, or develops diarrhea, it's time to push the panic button and call your local extension poultry specialist for advice.

WHAT AND HOW TO FEED

Just before a baby bird hatches and breaks out of its shell, the remaining yolk of the egg is ingested. This nourishment enables a newborn duckling to live without additional food for up to 72 hours after hatching. Even so, the sooner you give food and water to a baby bird, the better its chances for survival.

Start with a Duck Starter Ration

To start out, you should feed a complete duck starter ration of 20 to 22 percent protein. The ration should be in the form of crumbles or very small pellets. There is less waste when pellets are fed.

If you can't obtain duck starter at your local farm store or feed mill, you can feed them a chick starter ration of the same protein content. Try to buy unmedicated chick starter feed. Chickens are very susceptible to the disease coccidiosis, and their feeds often contain a coccidiostat. This medication can be harmful or even fatal to baby ducklings, who don't require this protection. Ducklings do require more niacin in their diet, up to 2 or 3 times more. If you are feeding a chick starter ration low in niacin, buy this vitamin in tablet form at a drug store and add it to the ducklings' drinking water. Alternatively you can add 5 to 7½ pounds of livestock grade brewer's yeast to each 100 pounds of chicken feed to supply the extra niacin. This works out to 2 to 3 cups of brewer's yeast per 10 pounds of feed. Ducklings deprived of sufficient niacin can have weak legs and retarded growth.

For the first 2 weeks of the ducklings' lives, keep feed in their troughs at all times, but only fill the troughs about half full. This will help prevent waste. Always make sure that they have plenty of drinking water so that they don't choke on their dry feed.

Switch to a Grower Ration

When the ducklings are about 3 weeks old, switch them to a grower ration, which will have a protein content of 16 to 18 percent. If you are in a hurry to produce meat, keep the ducklings confined or semiconfined. The less they walk around, the quicker they will gain. You can feed this grower ration until the ducks reach the desired weight of about 7 pounds.

Don't switch feeds abruptly as this can cause severe digestive upsets in the birds. If you have starter ration left over, mix it in with the grower ration gradually, until it's all gone.

Supplementing the Rations

If you provide the ducklings with a yard or pasture, any nutritional gains will depend upon the type of vegetation present. If the yard is fairly barren, they will have sunshine, fresh air, and exercise — all benefits of no small value — but no additional nutrients. If the yard or range is flat and poorly drained, the birds can turn it into a mudhole in short order and pools of stagnant water can cause disease. If, however, the pasture contains short, succulent green grass and a sprinkling of weeds and their seeds, and all the accompanying insects, the ducklings can supplement their diet with these natural foods and eat from 10 to 20 percent less expensive commercial feed.

The bugs and insects, all those creepy-crawly forms of life that many of us find repugnant, are full of protein, and the ducklings will convert them into lean meat. Some duck growers even affix electric lights 18 to 20 inches above ground level in their duck yards, and turn them on at night to provide an inviting beacon for flying bugs, and subsequently, a feast for the ducks.

Fresh table scraps from your kitchen and vegetables or fruits culled from your garden or orchard are excellent supplements for the ducklings' diet. I suggest that when you feed this type of supplement, only supply an amount they can clean up within 15 or 20 minutes. The birds will ignore wilted vegetables anyhow. One way to keep fruits and vegetables appetizing is to cut them up and put them in a pan of fresh water. My own ducklings have never had any digestive problems from eating overripe fruits, provided they were not moldy.

It is important to keep in mind that if you are interested in getting your meat ducklings up to good slaughter weight in the quickest time possible, don't tamper with their diet too drastically. Prepackaged commercial poultry or duck feeds contain the correct amounts and proper balance of proteins, fats, fiber, carbohydrates, vitamins, minerals, and other additives to provide the ducklings with all of the elements they need for a complete ration, from day-old to slaughter weight.

You cannot turn young ducklings out on pasture, even when they are fairly well feathered out, at say, 4 weeks of age, and expect them to do well on just green grass or bugs or grains alone. At this tender age, their system isn't equipped to handle this type of diet. They may just maintain themselves or more likely, deteriorate in condition. They certainly won't prosper and gain the desired weight and size that you are looking for.

If the birds are totally confined, using White Pekins as an example, they will weigh about ⅔ pound each and will have consumed about ½ pound of feed at 1 week of age. When 2 weeks old, the ducklings will weigh about 1⅔ pound each and have eaten over 2 pounds of feed, cumulatively.

At 4 weeks of age, the ducklings will weigh about 4 pounds each and have consumed approximately 7¼ pounds of feed. This is an excellent conversion rate, less than 2 to 1. However, by the time a duckling has gained a weight of some 7 pounds, it will have consumed 18 to 21 pounds of commercial feed. If the duckling is allowed access to good pasture, it will take longer to reach the desired weight of 7 pounds, but will consume less of the expensive prepackaged feed and have a leaner carcass.

As you can see, once the bird is about a month old, the ratio of feed per pound of weight gain increases. As with all poultry, the older the bird gets, past a certain stage, the more feed it requires to put on weight. The cost of feed is the most expensive part of raising poultry.

MOLTING

If you keep your ducklings past the age of about 2 months, they may commence to molt, which is a natural process of shedding their old feathers and growing new ones. During the molting process, the birds cannot fly, having lost their flight feathers. The male birds, the drakes, lose their colorful plumage and are almost indistinguishable from the hens or females. This is nature's way of providing camouflage for vulnerable male birds.

The first year that I raised Rouen ducks, a breed in which the males have very bright and spectacular coloration, I was a novice and inexperienced about molting. I couldn't figure out what was going on in my duck yard. One day I had 20 drakes and 20 ducks and shortly thereafter—it seemed like almost the next day—I had 40 females. I didn't understand what had happened to all my good old boys, the drakes. About 6 weeks later, the drakes had their adult plumage. The next time it happened, I understood and was prepared.

The Trouble With Pinfeathers

Pinfeathers are the young, tiny, immature feathers that appear before the molt and subsequently grow into full-size feathers. The molt usually lasts from 4 to 6 weeks, and slaughtering and dressing out a bird when it is in molt can be troublesome. Also, during the molt, ducklings are not at their best for eating.

The solution to this problem is to full feed your ducklings and get them up to the desired weight within 7 to 9 weeks and then butcher them. If they are on pasture and gaining at a slower pace, wait until they are 12 to 14 weeks old and are in full feather and ready for slaugther.

DECIDING WHEN TO SLAUGHTER

When the ducklings are 7 weeks old, it is time to weigh 3 or 4 of them and determine if they are ready for slaughter. When catching ducklings, don't chase them wildly around their pen or yard as their legs and wings can be easily injured. Instead, corner them inside their house or else drive them into a V-shaped arrangement of boards about 3 feet high. Pick each one up as formerly described, with one hand grasping the duck's neck gently but firmly, and the other securing the flapping wings. When the duck has calmed down, hold it close to you, with one hand supporting its weight under the breast.

If most of the ducks weigh about 7 pounds, it's time to butcher them. If they are 6 pounds or less, keep them on the same feed and care for another week or so until they gain the desired weight.

If the ducks are underweight, examine them closely. If they have a lot of pinfeathers, it might be best to carry them on for another 3 to 4 weeks until their new feathers have grown out, and they are of good weight. Dressing them out will be much easier if they are not molting.

BEFORE SLAUGHTER

The night before they are to be butchered, catch at least 2 ducklings of good weight and place them in a box or crate in isolation from the rest of the flock. Plan to butcher at least 2 birds in your initial effort. If the first attempt isn't perfect, you can improve your technique on the second bird. It is best if the isolation box has slotted or wire flooring so that any manure can fall through the bottom and thus not soil their feathers. Supply the birds with plenty of water but no feed the night before they will be killed. This will make the dressing out process much cleaner and easier.

SLAUGHTER TIME

The old-fashioned way to slaughter a duck is to hold it over a chopping block. With its head and neck extended, cut off its head with an axe. Even if you provide a bucket or pan underneath the bird to catch the blood, this is a messy procedure at best. A better way is to tie or shackle the bird's feet with twine at a level convenient for you. Then grasp the bird's bill with one hand. With a firm pull to keep tension on its neck, and with a sharp knife, cut the jugular vein, on either the right or left side of its neck, just next to the head. Hold the head until the bird is bled out, with a suitable container just below to catch the blood.

Another method is to use a killing cone, a funnel-shaped device that holds the bird upside down, with its head and neck exposed to the knife. If you use a cone, you will still have hold the bird's bill and keep tension on its neck to facilitate the process of cutting the jugular vein. Be patient, wait until the bleeding stops, then remove the head, and the bird is ready for plucking.

Note: Some butchers leave the head on, after the bird is bled out, so that they can hold the bill and have better leverage when dipping the carcass in scalding water, prior to plucking.

PLUCKING THE FEATHERS

In order to pick the feathers and down, you will need a large container filled with water heated to a temperature of from 130 to 140 degrees F. A plastic or metal garbage can works just fine. You can add a small amount of laundry detergent to the hot water to aid the process.

Holding the duckling by its feet (and bill, if you've left the head on), slosh it up and down in the hot water for at least 2 minutes. You want the water to penetrate to the skin, which will loosen the feathers and make for easier picking. If the water is too hot or the bird is held under too long, the skin can begin to cook and become brown and discolored.

After the bird has been scalded, you can begin plucking. Some people start with the large wing feathers, but you can begin anywhere you wish. It is important that you pull the feathers out in the direction in which they are growing, that is, with the grain. If you go against the grain, the bird's skin can tear or break open. If the feathers don't come out easily, put the duckling back into the hot water for another minute or so and begin again.

When all of the feathers have been plucked, there will be a number of unsightly hairs (filament feathers) left on the carcass. These can be removed by singeing. Use a candle or propane torch or hold the bird over a gas flame on the stove, being careful not to burn your hands. If there are any pinfeathers on the carcass, they can be removed by squeezing them out with pressure from your thumb and a dull knife.

Troublesome small feathers, down, and hairs can be removed by using wax. First, the bird is rough picked, removing the bulk of the large feathers, including those of the wings and tail. Then the carcass is dipped into a container of melted wax. There are specially formulated waxes available from commercial suppliers. You can also use the paraffin that is found on your grocer's shelves. If you are only dressing a couple of birds at a time, melt a small amount of paraffin over low heat in a saucepan, place the carcass in a suitable receptacle, and pour the wax over the small feathers and down. Then dunk the bird into a pail of cold water to quickly harden the wax. When the wax has congealed to a firm finish, peel it off the carcass, and the small feathers and down should come along with it. The waxing process can be repeated if necessary.

SKINNING

If you are concerned about the cholesterol content of ducks, you can remove a great deal of fat by skinning the bird. It should be said that when skinned, the carcass is not too attractive, and, when roasted, the meat tends to be dry. When roasting a skinned duck, baste constantly to ensure proper moistness. This whole step of skinning is optional.

With a very sharp knife, cut the skin around the neck, the wings, and the first joint of the legs. Then slit the skin all the way down the backbone to the tail. Lay the bird on its back and cut around the vent in a circular fashion. Starting at the neck, slip the blade under the skin and begin separating it from muscle and flesh. One hand will guide the knife and the other will lift the skin. In some areas the skin will come off easily; in other places it will tend to stick to the meat and you must be careful not to remove too much flesh.

EVISCERATION (DRESSING THE DUCK)

Lay the carcass on its back on a firm surface, such as a counter top or cutting board. If you wish, you can put paper underneath the bird to absorb blood and other fluids.

With a sharp knife, remove the feet at the first joint. Then cut off the neck close to the body. Make a very *shallow* cut in the soft section between the breastbone and the vent, about 3 inches long. Just below the surface of the skin where you are cutting lie the intestines and all kinds of innards that you don't want to pierce with the knife blade; thus, the cut has to be quite shallow. It is better to cut too shallow than too deep.

If there is flesh or membrane remaining under your incision, and the intestines are not quite exposed, the opening can be enlarged by pulling it apart with your fingers. Your whole hand is going to have to go through this incision so that you can grasp and remove all the organs and entrails from the bird. Put your hand inside the carcass, take hold of the mass of internal organs and viscera, and with a firm pull (try not to squeeze) remove them from the body. The heart, gizzard, liver, and intestines should now be out of the bird. At this point, one end of the large intestine will be leading to and still attached to the vent at the rear end of the body. Don't cut into the vent, but cut around it in a circular fashion and remove it from the carcass, along with the rest of the intestine. Place this mass in a bucket. You will deal with it in a few minutes.

Reach up into the body cavity toward the neck and feel for the windpipe (trachea) and gullet (esophagus). They will be located next to the spinal column. With a firm pull, remove and discard them.

The lungs consist of spongy tissue and are located next to the ribs on both sides of the spine. They can be easily pulled out with your fingers. Turn the bird over and find the oil gland which is located just above the base of the tail. Remove this gland by making a deep cut around it. Then wash the carcass, inside and out, with cold water.

Whether you are going to eat the bird within a day or so, or freeze it, it should be thoroughly chilled as soon as possible after evisceration. Also, aging the bird for 12 to 24 hours will make the meat more tender. The duckling can be chilled by placing it in a refrigerator with a temperature of 34 to 38 degrees F. Or you can put it in a pail of ice water. Don't skimp on the ice cubes.

Now for the Giblets

The heart should be rinsed off and placed in a pan of cold water.

There will be a small green gland attached to the liver. This is the gall bladder, and it contains bile. Treat it with caution. Cut it off of the liver and don't be afraid to take a smidgen of the liver with it rather then break the gland. The phrase, "bitter as gall" is not an empty one; any meat the bile touches can be spoiled and made unfit for eating. Wash the liver off and place it in the cold water with the heart.

In order to clean the gizzard, make a shallow cut into the yellow colored part around the outer edges. Pull it apart with your fingers and discard the inner part which will contain partially ground up food and small particles of grit. Wash the gizzard off and place it in the cold water pan with the heart and liver. The giblets can be frozen along with the duck or eaten immediately.

PACKAGING AND FREEZING

If you have chilled the bird in ice water, hang it up and let it drain thoroughly for 20 to 30 minutes before packaging. For the meat to keep well, you want to have as little moisture as possible in the package.

Freezer bags can be obtained from farm and feed stores or even some large supermarkets. In order to retain good quality in the meat when freezing, the wrapper should fit closely to the carcass, and this means that you must remove as much air from the bag as possible. One way to do this is to insert an ordinary drinking straw into the package after the bird has been placed within and draw as much air out as you can. Then twist the neck of the wrapper and tie it off tightly. A well-packaged whole duckling can be kept for up to 6 months in the freezer. The giblets or a cut-up duckling, will keep well for 3 to 4 months in the freezer.

If you are not too pleased with the result of your first effort at dressing out a duck and packaging it, don't be discouraged. Plunge right in and repeat the process with the second duck awaiting in the holding pen. In time, you will become quite proficient. It is a good idea, if possible, to have an experienced person go through the whole process with you the first time, from slaughter to packaging.

It is important to dispose of offal and other unusable parts of the duckling in a place where family pets or predators cannot get to it. Either burn or bury the viscera in a pit at least 18 inches deep.

Dressing Percentage

When dressed out, a duckling will average about 70 percent of its live weight. Thus, a 7-pound bird after dressing will weigh about 5 pounds, with neck and giblets included. Fifteen ducklings will provide about 75 pounds of meat for you and your family.

FEATHERS

It takes about 5 ducklings to produce 1 pound of feathers. After you have processed the ducks, wash the plucked feathers in lukewarm water with a bit of detergent added and spread them out to dry. Duck feathers are used by bedding and clothing industries. Companies that buy feathers are located in most large cities. Your extension poultry specialist can help you find potential buyers.

Chapter 9

RAISING DUCKS FOR EGGS

Besides raising ducks for meat, some home duck producers raise them for their eggs.

In a previous chapter, I stated that Khaki Campbells can lay up to 300 eggs per year and that Indian Runners will lay up to 250 eggs per year. I must qualify these statements. These figures are based on providing expert management and ideal conditions. To maximize egg production, the birds are fed a high-protein ration of commercial feed, kept semiconfined, and exposed to 14 to 16 hours of light per day.

Providing enough hours of light is a critical variable here. When the bird's eye is exposed to sustained light, the pituitary gland is stimulated; in turn, this gland secretes hormones which activate the hen's ovary into laying eggs. This means that you must provide artificial light at certain times of the year for top egg production. Exactly how much light is needed is not certain. Recent tests conducted at Cornell University suggest that laying hens can achieve maximum production with only 10 hours of light in staggered periods. Conclusive results of this experiment are still pending.

When the females are fed a high-protein ration but are exposed only to the natural daylight hours in the fall, their egg production falls to about 200 eggs per year. And when they are fed only coarse grains and exposed to natural daylight hours, their egg production averages about 150 eggs per year.

As you can see, management and feeding make all of the difference. If your birds are fed only whole or cracked grains, and they are left out on range and given no artificial light during the fall and winter, their egg production can be as low as half of their projected optimum rate. This could be a blessing if you are worried about having too many eggs on hand.

It is not necessary to have drakes, or male ducks, run with the hens in order for them to lay eggs. The hens will produce all the eggs they are ever meant to lay without the presence of a male. In fact, if you are interested in attaining peak production of nonfertile eggs, having drakes around can interfere with the hens' rate of lay. At certain times of the year, the drakes will hassle and chase the ducks around in the attempt to mate. This can upset egg production. If you want fertile eggs, however, you must have drakes with the hens. (Fertile eggs don't keep quite as well as nonfertile ones.)

STARTING YOUR
EGG-LAYING DUCK FLOCK

Your first step is to decide which breed to raise. Then you must locate a source for this breed and place your order.

For egg production, the best breed to choose is either the Khaki Campbell or the Indian Runner. Khaki Campbells look more like conventional ducks. Their posture is more horizontal, and they do waddle. Indian Runners stand tall and move with agile grace.

Both breeds are very hardy and are excellent foragers. Neither breed is known for its maternal instincts; they don't take to setting eggs very well. After the eggs hatch, neither breed is inclined to brood the young ducklings. Both breeds lay large quantities of white or white-tinted eggs, with Khaki Campbells being better producers, as far as quantity is concerned.

BUYING THE DUCKLINGS

For your first venture in raising ducks for egg production, I suggest that you buy day-old ducklings, rather than incubate eggs. Although it is sometimes possible to purchase fertile eggs from a hatchery, there will be no guarantee as to the success of the hatch. And the hatching rate can be quite low, as low as 60 percent live birds, or less. The eggs will cost about $1 apiece. Unless you pick them up locally, there will be transportation charges.

Buying day-old ducklings is more economical.

Suppose that you want to raise 10 hens of a high-producing egg breed, assuming that this number of hens will supply your family with sufficient eggs for home consumption. Given a 60 percent projection of a successful hatch, you will have to purchase about 34 fertile eggs to come up with 20 live ducklings. Assuming that half of these will be females and half would be males, according to normal sex ratios, you will then have 10 potential egg layers. (In general, the ratio of males to females hatching out from eggs is 50–50. That is, out of 100 births, there should be 50 drakes and 50 hens born. With lesser amounts, this ratio does not necessarily hold true.)

The fertile eggs will cost at least $34, and you will have to buy an artificial incubator or else try to find about 7 broody hens, at about $2 each, to set the eggs for some 28 days.

Estimated cost of incubation:

34 fertile eggs	$34.00
Electric incubator	35.00
	$69.00 Total

Compare this with buying 20 day-old ducklings of the same breeds, which will cost about $35, and you don't have to wait for 28 days. Also, they are guaranteed 100 percent live arrival if purchased from a reputable hatchery. Day-old ducklings are usually sold as-hatched, or unsexed; thus, out of 20 birds, everything being relative, you should get 10 hens and 10 drakes.

Clearly, if you are just looking to start a flock of egg producers, it makes the most sense economically to start out with day-old ducklings. If, on the other hand, you are going into the duck breeding business, then you will want to purchase incubators anyhow. In that case, provided you can find a good source of fertile eggs, you may want to start out with eggs.

HOW MANY DUCKLINGS TO BUY

I suggest purchasing 20 day-old ducklings. Half of them will be hens and the other half drakes. (It is very difficult to find sources of sexed ducklings, and they are usually half again as expensive as straight run, or unsexed ducklings.) You keep the females (hens) for egg laying, which they should begin at about 5 months of age. You can butcher the males when they are about 2 months old and weigh 3½ to 4 pounds. They will be lightweight roasters, but will be lean and taste delicious.

WHERE TO ORDER YOUR DUCKLINGS

Again, buy from a reputable hatchery. Ask your extension poultry specialist for reliable sources, or consult local duck growers. Use the listings provided by the National Poultry Improvement Plan catalog (see appendix for an address).

ORDER YOUR DUCKLINGS

Late spring or early summer is the best time to have your ducklings arrive. Ducklings purchased at this time will begin laying eggs by fall. Starting off the flock when the weather is temperate will help you in your first duck growing venture. Be sure to place your order at least 1 month before you want them to arrive.

THE DUCK HOUSE

With the ducklings on order, it is time to get their housing and equipment needs in order. For the first 8 weeks of their lives, egg-producing ducklings have the same housing and equipment requirements as meat ducks. After that you must provide additional space for the mature ducks and larger capacity feeding and watering units, as well as nests. While you may have meat ducks for just a couple of months before they are slaughtered, these birds you are raising for egg production or to keep as potential breeders may be with you for a couple of years or more.

Although young ducks need only 2½ square feet of space, mature ducks will require 5 to 6 square feet of space when confined. Thus 10 birds will need 50 to 60 square feet of space. But you must also provide space for the nest boxes. It's also a good idea to leave space to store feed and other supplies in one corner of the duck house.

Even if you plan to raise your ducks partially on pasture, you should plan to keep the ducks confined at night for several reasons. First, it is important to protect ducks from predators. Second, in a duck house you can provide them with supplemental artificial light during the fall and winter to maintain their egg production. Finally, duck hens lay their eggs at night, or before 7 or 8 AM, and they aren't too fussy about where they lay them — including in the driveway, in a pool of water, or out behind the barn. If the ducks are confined within a shelter at night, then turned out when the sun is high the next morning, you will have a much better chance of collecting usable, edible eggs. And that's the main idea.

The floor of the duck house can be made of dirt, although wood or cement is easier to keep clean. A layer of litter on the floor will keep the ducks warmer and will absorb moisture. The litter can be removed periodically and composted. The composted litter can be added to the garden, providing a rich source of nitrogen.

The doors and windows of the duck house must close tight to protect the ducklings from predators. It is a good idea to cover the inside of windows with a fine wire mesh to prevent wild birds from flying inside.

The duck house should be wired for electricity. Although mature birds do not need supplemental heat within the house, even in the coldest weather, the young ducklings do. Also, you will need the electricity to provide supplemental light for the laying birds.

Although the mature ducks are well-insulated from cold, they should be protected from drafts.

There is no reason you cannot use the same duck house from start to finish, for brooding the ducklings and for housing laying birds, as long as you provide enough space for the mature birds.

The Duck Yard

As mentioned earlier, you will want to provide the ducks with access to the outdoors. The ideal duck yard allows 10 to 25 square feet of ground space per duck, a source of shade, and a source of drinking water. The ground should drain well and be free from standing pools of water.

Shade can be provided by trees, shrubs, A-frames, or rough shelters made from scrap lumber.

Fencing is a good idea, particularly where there are predators. The fence should be of woven wire, about 5 feet high, with 1 strand of barbed wire 2 inches above ground level outside the fence and a strand of barbed wire at the top of the fence.

PONDS AND STREAMS

A duck takes to water like a . . . duck. Well, why not? They are waterfowl and don't even have to be taught to swim. They look very picturesque when exercising their natural talent on a pond or stream.

Ducks like to feed on aquatic plant life and little fish. These things are good for them. However, too much fish in their diet will give their meat or eggs a fishy taste. Although some commercial rations contain fish meal, the proportions are carefully worked out so as not to cause fishy flavor. And, on big ponds or lakes, snapping turtles and large fish can be a threat as they are quite fond of duckling.

In any case, ducks can be raised just fine without them ever swimming a single stroke on a body of water. They must, however, always have a good supply of drinking water. If you do have a pond or stream, there is no reason your ducks cannot enjoy swimming on it. It will be fairly easy to entice them back to land at feeding time. After their evening meal, they could be penned inside their house for the night.

A spring-fed pond or moving stream is one thing, but a pool of stagnant water within the duck yard can cause disease and is a no-no!

NECESSARY EQUIPMENT

When the ducklings are young, their equipment needs are the same as for meat-type ducks. You will need to provide a guard, or corral, to keep the ducklings confined to a small area in which you will locate a heat source, waterer, and feeder. The heat source provided is usually a 250-watt heat lamp. For more details on the equipment needed during brooding, see chapter 8.

Waterers

Always supply plenty of drinking water to the ducks at all times. They use a lot more water when feeding than chickens do. You can use fountains, waterers, buckets, pails, or old pots and pans. The containers should be deep enough—about 3 inches deep—so that the birds can rinse their bills, nostrils, and, on occasion, even their eyes in the water. Don't use a container that a bird could get caught or stuck in and thus drown.

If the birds are confined in a house, do not allow them access to feed unless they also have water. They can choke on dry feed.

Ducks are messy eaters and if their waterers are inside a house, they will make the surrounding area quite wet. One way to remedy this is to place the waterers on a stand low enough so that the ducks can climb up on it to drink.

The stand is easily made. Nail together 1 x 4 boards, each 2 feet long, so that they stand 4 inches high in the form of a square. Affix a 2-foot square piece of sturdy wire mesh or hardware cloth on top of the boards. That's all there is to it. Place the water fountain on top of the wire in the center. If the birds have trouble climbing up, make a small ramp for them, or else bank the floor litter higher on the outside of the stand to give them easy access. If you don't use a stand,

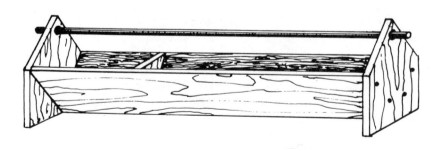

A homemade feeder can be divided into sections—one part for feed, the other for grit.

remove the wet litter from around the waterer regularly and replace with dry bedding.

If you started out with just 1 waterer for the brooding ducklings, you will need to add a second waterer by the time the ducklings are 8 weeks old.

Feeders

The same chick feeders that are used for brooding meat ducklings can be used for brooding egg-producing ducklings. (See page 43.)

Feed troughs are usually provided for older ducks. Allow 6 linear inches of space per bird. If the trough is designed so the ducks can eat from both sides, a trough 5 feet long should be sufficient for a flock of 10 ducks. A spinner bar at the top will prevent the ducks from climbing into the trough and soiling the feed.

Nest Boxes

When the birds begin to lay, you should provide nest boxes for them. If you don't provide nests, they will lay their eggs any old place, making their own nests in the floor litter, which some of them will do anyhow.

One nest should be provided for every 4 to 5 laying birds. The nests can be made of scrap lumber. Each nest should measure 12 inches wide, 12 inches high, and 16 to 18 inches deep. The front of the nest should have a board about 2 inches high nailed to the opening. This short front board serves to keep clean litter inside the nest and prevents the eggs from rolling out. The top and front of the nest will be open. If the partition boards are placed against one wall of the pen, this will close off the back of the nest.

The nest should have a couple of inches of clean litter within.

Nest boxes are set on the floor in the duck house. Nest cube can be made 12" to 24" square, depending on size of ducks.

WHEN THE DUCKLINGS ARRIVE

Before the ducklings arrive be sure the duck house is thoroughly cleaned and that your heat lamp is functioning properly. You will need to provide the newly arrived ducklings with feed, water, and a comfortable 90 degree F. space. See pages 117–120 on preparing the duck pen and brooding newly arrived ducklings. For the first 8 weeks of their lives, egg producing ducklings have the exact same care requirements as meat ducklings.

FEEDING

For the first 2 weeks, the ducklings should be full fed a duck starter ration of 20 to 22 percent protein. Keep feed and water in front of them at all times. If duck feed is not available at your local farm store, an unmedicated chick starter ration can be substituted. You may have to supply extra amounts of niacin, which can be obtained at any drug store. Add the niacin supplements to their drinking water. Or supplement the ducks diet with brewer's yeast as described on page 120. If you do use chick starter ration, check it out with your local extension poultry specialist. He or she is there to help you, and the ones I have been in contact with have always been helpful.

Next, Feed a Growing Ration

From the time they are 3 to 8 weeks old, the ducklings should be fed a growing ration, which will contain 16 to 18 percent protein. Again, keep feed accessible to the birds at all times. Always try to change feeds gradually. If you have starter ration left over, mix it in with the grower ration.

Switch to a Breeder-Developer Ration

When the ducklings are 8 weeks old, they should be switched to what is termed a breeder-developer ration. This feed will have a protein level of 15 to 16 percent and contain less energy (fewer calories) than the starter or grower feed. The point here is you are not trying to get these ducklings up to a heavy weight in the quickest time. These hens are not destined for slaughter. You want to maintain them in good condition and have them gain weight slowly.

If the females are fed a ration too high in protein or energy at this stage of their lives, they can come into egg production too soon. This can result in small and fewer eggs, and also cause internal problems with the hens.

Light breeds of ducks, such as Khaki Campbell and the Indian Runner, can begin laying eggs when they are about 4½ months old. It would be just as well if they didn't start laying until they were at least 5 months old (20 weeks). Heavier, meat-type breeds, such as the White Pekin, Aylesbury, or Rouen, will usually start laying when they are 6½ or 7 months old.

The birds should be fed this breeder-developer ration until shortly before they are expected to begin laying eggs. For example, if you are raising Khaki Campbells or Indian Runners, feed them the breeder-developer ration until they are about 18 weeks old. At that time, gradually switch to an egg layer ration of 16 percent protein, but fewer calories.

The laying ration will contain more brewer's yeast, vitamin A, and soybean oil meal and less wheat and alfalfa meal. Feeding this ration to prospective layers for 2 or 3 weeks before they actually come into production should put them in top condition when they get down to business. (Later on in the year, when the hens are not producing, due to molt or other reasons, they can be maintained on a less expensive feed ration of 13 to 14 percent protein.)

Substituting Grains for Commercial Feed

If you have grains available, or are raising a large amount of ducks and feel that it would be more economical to supplement expensive prepackaged feed with grain, a suitable ratio would be half and half. As an example, for every 100 pounds of feed, 50 pounds could be layer ration and 50 pounds could be grain. A suitable grain mixture consists of 3 parts corn, 1 part wheat, and 1 part barley. (The 50 pounds of grain would thus include 30 pounds of corn, 10 pounds of wheat, and 10 pounds·of barley by weight.)

Other grains can be substituted if done so in correct proportions. Remember, you are striving for a ration of 16 percent protein for your layers and just any old mix will not do. Consult your extension poultry specialist for a formula that will give you the appropriate amounts if you are going to mix commercial feed with grains.

A typical duck starter ration, complete with all the ingredients necessary for their growth and health, would include the following: yellow cornmeal, soybean oil meal, alfalfa meal, fish meal, dried whey, dried grain distiller's solubles, di-calcium phosphate, ground limestone, salt, manganese sulfate, zinc oxide, D.L.-methionine, vitamin A, vitamin D3, vitamin K, riboflavin, niacin, vitamin B12, calcium pantothenate, and choline chloride—and all mixed in correct proportions. A batch of this would be pretty hard to mix up in the backyard. If you're only raising a small flock, it would not be economical to do so.

The aforementioned ration is for ducklings from day-old to 2 weeks of age. As the birds get older and grow in size and weight, their needs change. Some proportions of the basic elements increase or decrease, some new ingredients are added and some deleted.

Alright, you may ask, how come ducks have existed for thousands of years without requiring such complicated rations?

Well, the wild duck got its food from the sea, tideland, marsh, and fields, and fulfilled all of its dietary needs from natural feeds. The wild Mallard, probably the common ancestor of all our domestic ducks, excepting the Muscovy, laid a clutch of some 6 to 12 eggs, once a year, in early spring. It took one of the resulting ducklings the rest of the spring and all summer long to gain a weight of perhaps 3 pounds or so.

Contrast this with today's White Pekin duck, which can gain 7 pounds in 7 weeks under expert management. Or the Khaki Campbell, which can lay 300 or more eggs per year. Unless you are raising ducks for fun or as a hobby, 12 eggs per year or a 3-pound meat bird just won't cut the mustard.

We push our domestic ducks for commercial production, whether it be for eggs or meat. I can sympathize with commercial duck growers who turn their birds into machines in order to make a profit, considering their large investment of time, money, and equipment.

Personally, I favor a middle route. Get the ducklings off to a good start with a complete ration. Then let them forage on natural foods in pasture, with a supplement of commercial prepackaged rations and grain. It will take a little longer to get the desired results. The resulting meat will be leaner, and you will get fewer eggs. Unless you have a ready market, it's hard to dispose of surplus duck eggs, anyhow.

Substituting Chicken Feed

If your local farm and feed store does not carry the specified duck starter, grower, developer, or laying rations, chicken feeds can be substituted. Again, consult your extension poultry specialist for advice. If you do feed chick rations, try to avoid medicated feeds. You may have to add niacin to their drinking water.

One drawback to feeding chicken rations is that they often are packaged in mash form. Pellets or crumbles are better suited for ducks, as they will waste a lot of mash due to their feeding habits. If you feed your ducks outdoors, mash can be gone with the wind, while pellets will not. If feeding on the ground, distribute the feed widely, so that all the birds can get their share, and timid hens won't be kept from eating by the bullies. If you are using troughs, allow at least 6 inches of space per bird.

Keeping Laying Hens

Laying hens will consume .5 to .6 pounds of feed per day, each. You are not going to full feed them as you don't want them to become too fat, just keep them in good condition. If you are raising 10 hens, as suggested, measure out 5 pounds of feed each day, and feed half of it in the morning and the other half at night.

Insoluble hen-sized grit and ground limestone or oyster shell should be provided, free choice, for the birds at all times. The grit will lodge in their gizzards and help them grind up feed. The limestone or oyster shell is to replace the calcium they are expending when laying eggs. The grit, limestone, or shell can be sprinkled lightly on top of their feed or placed in separate troughs or containers.

THE IMPORTANCE OF PASTURE

It is recommended that when ducks are provided with a yard, each bird be allowed a minimum of 75 square feet. The 10 layers you are keeping will need at least 750 square feet. That's really not a very big area. But, any size yard is better than no yard at all. Again, it depends on the terrain and type of vegetation.

While ducks are not as good foragers as geese, they can supplement their feed requirements by 10 to 20 percent if pastured on a well-drained range of short, green grass.

Grass will supply protein, calcium and manganese. It will supply more vitamin A than the ducks need, whether they are still growing or already laying eggs. Grass also supplies all the riboflavin ducks require. It will not supply vitamin D; but when the ducks are out on pasture, the sun will provide all of that vitamin they need.

If a mature hen eats about 200 pounds of feed per year, and you are buying prepackaged commercial feed priced at about $11 per 100 pounds, you are spending $22 per year per hen in feed costs. If that duck can supplement her diet to the tune of 20 percent on grass pasture, that's a savings of $4.40. Multiply that by 10 hens and it's a savings of $44. Not insignificant.

DETERMINING THE SEX OF DUCKLINGS

Somewhere along the way, if you purchased 20 unsexed ducklings, you are going to want to separate the hens from the drakes in order to know which birds to butcher and which to keep for egg layers. In order to do this it is necessary to examine each duckling individually. Caution: if you don't remember how to catch a duck without injuring it, please refer back to page 106.

Feather Coloring

The easiest method to determine a duck's sex is by its feather color. But by no means is this a foolproof method. In fact, it only works with breeds having colored plumage, such as the Rouen. The feathers on the backs of the males will be darker than those of the females by the time they are both 8 weeks old. (The males won't have their unmistakable, striking feather coloring until they are about 5 or 6 months old.) With breeds like White Pekins or Aylesburys, it is hardly possible to distinguish between the sexes by feather color.

Body Size and Feathers

Adult males have larger heads and bodies than females, and they weigh a pound or 2 more. But when the ducks are still immature, this method is pretty much a guessing game. Drakes have curly tail feathers, but this trait usually doesn't show up until they are fully grown.

Voice

Catch and hold each bird individually. The females will protest with a loud, strident quacking. The males will speak in a low, throaty voice. It should be noted that this system won't work very well until the birds are 7 to 8 weeks old.

Examining the Reproductive Organs

The only fool-proof method to determine the sex of the birds is to examine the reproductive organs. Under a bright light, the duck is held on its back with the head towards you and the tail pointed away. The tail is then bent downwards and a finger is inserted through the vent (external opening) and into the cloaca, the chamber which contains the reproductive organs. Gentle pressure is exerted from around the outside of the vent until the opening is enlarged and the genital organs can be seen.

If the duck is a male, a tiny penis will be visible near the center of the cloaca.

Young ducklings can be easily injured by a fumbling novice, and there is a lot of room for error in this process. The best course is to have a knowledgeable person demonstrate the whole examining process for you.

BUTCHERING THE DRAKES

Males of heavier breeds, such as White Pekins, should be ready to butcher after 8 weeks of eating well. Weigh several drakes. If they average about 7 pounds per bird proceed to prepare them for the freezer.

With breeds such as Khaki Campbells or Indian Runners, the males should weigh 4 to 5 pounds before butchering. If they have not reached this weight by 8 weeks of age, you may have to carry them for another 2 to 4 weeks. Mark the drakes with metal or plastic leg bands or spray paint, so you don't have to go through the whole sexing process again at a later date. Or place the drakes in a separate pen, with another feed trough and waterer.

The young drakes will make excellent lightweight roasters. The butchering process is the same as for meat ducks. See pages 123 to 126 for details.

EGG PRODUCTION

Now that you have spent about 5 months in caring for the hens and seeing to their needs, it's time for them to do something for you and they will.

Supplemental Lighting

If you want to increase your chances for good egg production, and your hens are approaching their egg-laying stage in the fall when the hours of natural daylight are decreasing, begin a program of supplying artificial light in the duck house. Begin with the light about 2 or 3 weeks before the hens are expected to start laying.

The idea is to provide the birds with at least 14 hours of continuous light. One 40-watt bulb is sufficient to light an area of several hundred square feet. You can turn it on before first light, or before sunset, or a combination of both. If you don't like to get up early, you can even leave it on all night long, with no harm done. But, you want to come up with a minimum of 14 consecutive hours of light. You can buy an automatic timing device which will do the job and save you the early and late trips to the duck house.

THE EGG PRODUCTION CYCLE

The first eggs will be small, and the hens may not be consistent in their laying, but egg production will increase rapidly. It would not be uncommon if within 6 weeks or so after the hens begin laying, the flock swings into full production — with each of your 10 hens laying an egg every day. The hens usually remain in high production for 5 or 6 months. After that, their production may fall off to less than 50 percent.

When they molt, the hens will stop laying altogether. Molting is a natural process where birds shed their feathers and grow new ones. The molt can last from 6 to 8 weeks. Then the hens will begin laying again. Some commercial growers bring the birds into a forced molt ahead of their natural time, and then get them back into production as soon as possible.

Ducks will lay the most eggs during their first year of production. But unlike chickens, whose production decreases markedly in their second year, there is usually only a minor decrease in the ducks' production during the second year. The average productive life of a chicken is figured at 1 to 2 years. A duck can produce well for 2 to 3 years. Some ducks lay for 6 to 8 years.

Because of their marvelous biological make-up, it takes a little less than 24 hours for a duck to cycle an egg. With the chicken, it takes something more than 24 hours.

FERTILE EGGS

If you decide to hatch out some eggs, you must have a drake in with the hens to produce fertile eggs. One drake to every 4 or 5 hens is considered adequate. Select only the cleanest, unbroken eggs for hatching purposes. Eggs for hatching can be stored for up to 10 days with good results before they are to be set or incubated. They should be stored at a temperature of about 55 degrees F., large end up.

DISEASE

Next to geese, ducks are about the most hardy of all poultry. Prevention of disease and sanitation go hand in hand. If you pick up the soiled or damp bedding in their house and replace it with fresh, dry litter daily, there should be no problems in a small flock.

If you supply plenty of water and feed the correct ration, the chances of your small flock of ducks becoming ill are nil. Observation is the key. If a duck doesn't eat or drink and has a watery discharge, isolate it from the rest of the flock. If more than 1 bird is affected, consult your local extension poultry specialist for advice and treatment. Any dead birds should be buried in a pit at least 18 inches deep or else burned.

The duck yard should be well-drained and contain no pools of stagnant water.

RAISING DIFFERENT TYPES
OF POULTRY TOGETHER

I have raised ducklings and goslings together with good results. If ample water and feed were supplied for all the birds, there was no problem. The only hitch occurred when the birds reached mating age. At that stage, the geese, being larger and more aggressive, bullied the ducks, and I had to separate ganders from drakes, in order to save the latter from being injured. In any case, poultry of different ages should never be raised together and subfamilies should be raised apart, if possible.

PREDATORS

Crows, ravens, hawks, and owls all have a sweet tooth for young ducklings. If the ducklings are confined to a shelter for at least the first month of their lives, they will be too big for the first 3 avian predators to handle; and if they are penned in at night, owls can't get at them. As suggested, a tight and secure duck house will keep out weasels, mink, raccoons, skunks, foxes, dogs, and cats.

MANURE

Manure and litter gathered from the duck house are an excellent addition to your compost pile. Fresh duck droppings contain a high percentage of nitrogen and are considered a "hot manure"; direct application of fresh manure is not recommended for gardens and lawns. Let the manure age for several months, then use it as a fertilizer.

Part Three: Geese

Chapter 10

WHY RAISE GEESE ?

Geese have been around for a long time. Any creature that has contributed so much to myth and folklore and enlivened our daily speech with so many colorful expressions has to be pretty ancient.

We have Mother Goose, Father Goose, the Goose That Laid the Golden Egg, the Spruce Goose, Silly Goose, and the Goose Girl. Not to mention goose bumps, gooseneck and goose step.

In the current craze for trivia games, one best-selling game poses the question, "What bird has been domesticated by man for the longest time?" The answer is the goose, first domesticated by man about 4,000 years ago. Fads come and fads go; by next year, trivia games may be as stale as yesterday's news, but the goose will survive.

In the United States, it is estimated that only one million geese or so are produced annually. However, much more are produced and eaten in Europe, Asia, and Africa, where it all began. The Egyptians considered the goose a sacred bird. Sacred or no, geese make darn good eating, which is only one of many reasons to raise them.

SIX REASONS TO RAISE GEESE

The most obvious reason to raise geese is for meat. Geese provide delicious eating. When roasted, that crinkly crunchy, golden brown skin covers a delectable feast.

Geese also provide eggs for eating. It is not unusual for a goose to be a productive egg layer for up to 10 years. Their eggs are quite large in comparision with chicken eggs, and most people use them only for baking or cooking. One of my geese once laid an egg that weighed 7½ ounces. I sent the good news to a local newspaper. Unfortunately, an ostrich egg beat me out and I didn't make the *Guiness Book of Records*. I fry, poach, and use goose eggs in omelets, but the white of the egg can be quite coarse and grainy.

However, it should be said that the best laying breeds of geese can produce only about 25 percent of the annual egg production of Leghorn chickens or Khaki Campbell ducks. A small flock raiser shouldn't keep geese for high egg production.

We also raise geese for fun, or as a hobby, or for show purposes. They make great conversation pieces. A White Chinese goose, in particular, with its swanlike neck and graceful carriage, is a handsome sight when it walks through a patch of green foliage or swims on a pond.

Geese are very inquistive and aggressive, making better watchdogs than most German shepherds, always raising a ruckus when a stranger pulls into the yard. Although geese are called honkers, their call has always reminded me of a pack of baying hounds or a squeaking rusty hinge on a heavy iron gate.

Young geese can be put to work as weeders in cotton fields, orchards, nurseries, and strawberry patches, taking the place of expensive hand labor. It is said that 8 geese can replace a man with a hoe. They don't get stuck in the mud or injure the roots of plants as heavy machinery can do.

Geese also provide valuable feathers and down, which can be made into warm clothing and bedding.

Why raise geese? Because they are ornamental as well as a good food source of meat. These beautiful birds are Canada geese. (Photo courtesy USDA)

WHAT'S INVOLVED WITH RAISING GEESE?

It has been my experience that geese are the most hardy of all domestic birds. I can't remember ever losing an adult goose to illness or disease. It is not unusual for a goose to live for 30 years.

You don't need fancy buildings to house them in. They do need a snug, draftfree shelter for at least the first month of their lives. After that, they usually prefer to be outside. Don't forget, however, that geese are noisy; you can't keep them in the middle of town. Check with your town clerk about local ordinances before you bring any geese home.

It is not at all necessary to have a pond or stream in order to raise geese successfully. They must, however, always have a plentiful supply of drinking water. They are much more at home on land than their cousins, the swan or duck.

The basic natural diet of geese is grass. Grass can provide a complete diet for geese throughout most of their life cycle. With the exception of young goslings during the first few weeks of their life and breeding females during their period of egg production (when the birds do require some grain), all they need is good pasture or range.

I do suggest that you *do not* attempt to raise geese if they will not have access to a pasture, range, or large yard. It wouldn't be fair to the geese and it wouldn't be much fun for you to keep an indoor area, occupied by totally confined adult geese, reasonably clean and dry.

Geese raised partially on pasture will be much leaner than geese raised in confinement, which is how most commercial geese are raised. In fact, when people complain that geese are too fatty, or give us too much cholesterol, they are not referring to range-fed geese that have foraged on grass, weeds, insects, and such. Your geese, raised on pasture, will be leaner than the storebought geese.

When raised for meat, geese show a rate of gain that is fabulous, and they have no rivals. A Rock-Cornish broiler chicken can weigh 4 pounds when 8 weeks old. A White Pekin duckling can weigh 7 pounds at the same age. But an Embden goose can weigh 10 pounds at 8 weeks of age and be only partly grown. This figure applies to birds that are full fed and under expert management. The chicken and the duck are ready for the table at those weights. The goose has some more growing to do. When a goose is 10 to 12 weeks old, its rate of gain decreases.

Despite their remarkable rate of gain, you will not save any money by raising geese for eggs or meat. A frozen goose at the supermarket will weigh an average of 9 pounds and cost a little less than $2 per pound, or close to $18.

A one-day-old gosling can cost from $5 to $8 depending on the breed purchased, which is much more expensive than chicks or ducklings. I recommend that you start with 8 goslings. They will eat a commercial ration for 30 days, then forage for themselves for 2 or more months. In order to reach a live weight of 10 to 12 pounds in 12 to 14 weeks, these 8 geese will consume about 300 pounds of commercial rations, which will cost around $40.

Although housing and equipment costs are negligible, if you have to go out and buy new feed troughs, waterers, and a brooder lamp, you will spend about $20. Then there is the big expense: fencing. Depending on the type of terrain you are enclosing, as well as its size, this expense can exceed $100.

In regard to the time involved, from start to finish, if you are raising geese for meat alone, you will probably keep them for 3 to 5 months, depending on how they are fed and managed. If you are raising them for other reasons, they may be your interesting companions for 10 years.

If you decide to raise geese, you or someone must be on hand to feed, water, and care for them every day, 7 days a week. It is not a time-consuming job, probably requiring only 15 minutes twice a day, but you can't go away on vacation and leave them alone.

It will be best if you learn how to butcher the birds yourself, and prepare them for freezing. And, you should have a large freezer for storing them.

Is raising geese worth the expense and effort? Well, a large portion of that question can be answered by a simple taste test. If you've never tried roast goose, go out and buy a goose and sample it. It would also be a good idea to sample a goose egg, as well.

If you are lucky enough to find someone in your area who raises geese, it might be possible to purchase a couple of fresh goose eggs from them and try them out at home. Or ask your extension poultry specialist or at your farm and feed store for names and addresses of people who raise geese. If possible, buy nonfertile eggs.

Be prepared for a surprise. Accustomed as you are to the size of the large (chicken) eggs you bring home from the store, the goose eggs may amaze you. One grade A large chicken egg will weigh about 2 ounces. A goose egg will weigh 3 to 5 ounces more.

If you are able to purchase fresh goose eggs, ask the people you buy them from how they prepare theirs for the table. Again, keep in mind that geese are raised primarily for meat and not egg production.

Having tried roast goose and enjoyed it, as I am sure you will, if you decide to raise a small flock, have a building of some sort and a bit of land for them to range on, and are willing to devote a small but consistent amount of time to the project, the next step is to examine the breeds.

Chapter 11

BREEDS
OF
GEESE

If you've ever eaten a goose from a supermarket or restaurant, chances are you have eaten an Embden or Embden-White Chinese cross. It was most likely about 14 weeks old when slaughtered. The Embden or Embden-White Chinese cross is one of the fastest growing varieties and can reach a live weight of from 12 to 13 pounds in 14 weeks. The bird was probably raised in confinement or semiconfinement and full fed in order to gain that much weight in such a short time. Time is money to the commercial goose grower, who requires a fast growing bird with white feathers.

Birds with white feathers, such as the Embden and Chinese, are popular with commercial growers because they tend to yield attractive, blemish-free carcasses when plucked. Because of their white plumage, any existing pinfeathers are also white and don't leave any dark blotches in the skin.

The geese you raise for your own use can be any color or multi-colored. Their carcasses don't have to be perfect when dressed out and plucked. They won't be on display in a brightly lighted case or poked about by finicky consumers.

For a home flock, I always recommend raising geese partially on range, letting them forage on grass and insects. A leaner carcass is desirable to me, because the meat will be less greasy.

The swanlike appearance of White Chinese geese belies their aggressive nature. These geese are noted for their high egg production.

COMMON GOOSE BREEDS

Here's a rundown on the most commonly available goose breeds.

White Chinese

A most attractive white-feathered bird, with its long swanlike neck and graceful carriage, a White Chinese goose is smaller than most other geese. White Chinese geese mature rapidly and are the best of weeders. Inquisitive and very aggressive, they can be fed twice a day, every day, the year around by the same person and they will still stretch out their long sinuous necks, crouch low to the ground, and hiss threateningly every time their friend approaches. The threat is pretty much pure bluff, but I appreciate their aggressive nature. It helps ward off predators. One winter bloodthirsty weasels killed off all my chickens and ducks but didn't touch one White Chinese goose.

The adult male will weigh 12 to 14 pounds. The females are not very motherly. They don't like to set eggs or brood goslings, but they are about the best egg layers among all the breeds of geese. A hen can lay approximately 40 to 65 eggs annually.

There is another variety, the Brown Chinese, with all of the same qualities. They are not as popular as the Whites.

Embden

This large white-feathered goose is probably the fastest growing bird among all the breeds; it can gain 10 pounds in 8 weeks. It is a favorite with commercial growers. Embdens are not as aggressive as White Chinese. An adult male will weigh 26 to 30 pounds. The females are very good mothers and will lay 30 to 40 eggs annually.

African

This is a spectacular bird, with a large black knob on the top of its head, a black bill, brown feathering down the back of its neck, and various shades of brown and gray coloring on the rest of its body. Africans grow quickly, mature rapidly, but are not popular with commercial growers as a market bird because of their colored feathers (and potential dark pinfeathers). (White Feathers have a higher market value.) A mature male will weigh 18 to 20 pounds. The female will produce about 30 to 40 eggs per year.

Pilgrim

This docile, medium-size bird comes in 2 colors. It is the only breed that can be sexed by color at 1 day of age. When hatched, the males are a creamy white and the females are greenish-gray. When fully grown, the males are all white with blue eyes, and the females are a handsome gray with hazel eyes. An adult gander will weigh 14 to 16 pounds, and, like any good liberated male, will help his mate in raising the goslings, if they are from a natural hatch, (not artifically incubated). The females will each lay 35 to 45 eggs per year and are good at setting eggs and brooding their young.

The Pilgrim is a medium-size bird that is excellent for meat production. The ganders are a creamy white and the females are gray and white. (Photo courtesy USDA)

Toulouse

One of the larger breeds, Toulouse geese are broad and deep-bodied. They look bigger than the Embdens because they are loose feathered. Their plumage is a combination of dark gray on the back, a lighter gray on the breast, and a white abdomen.

The name is derived from the city located in southern France in an area noted for its geese. French farmers constrict their necks with string or bands, feed them heavily, and the resulting paste made from their enlarged liver is sold to gourmets as pâté de foie gras. (Ugh, horrible process). The females are not considered good setters or brooders, but will lay 30 to 40 eggs annually. An adult male can weigh 26 to 30 pounds.

Sebastopol

A Sebastopol is a very attractive bird with long curved, plumelike white feathers on its back and sides and, short curled feathers on its lower parts. Although adult Sebastopols can reach a weight of 12 to 14 pounds, they are considered an ornamental bird and are a favorite with breed fanciers who exhibit at poultry shows.

Egyptian

This is a tall, long-legged, but very small goose. Its mature weight is only about 5½ pounds. Egyptians are raised for ornamental or exhibition purposes. Their coloration is varying shades of gray and black, with spots and splotches of red, brown, and white. I have only seen one of these birds in my life, and it dwelled all summer long with a large flock of wild Canada geese near a pond about 2 miles from where I live. It seemed quite at home on land; I never saw it go into the water or swim or even fly. When winter came, the wild Canada geese abandoned their consort and flew south. I'll never know the fate of that lonely Egyptian.

CHOOSING A BREED

Choosing which breed is best is really determined by what you want geese for. If you want a fast growing bird to put meat on your table, then Embden, White Chinese, or a cross between the two breeds is best. For egg production, the Chinese is best.

If you want to use geese as weeders—for orchard, nursery, vineyard, cottonfield, or strawberry patch—the White Chinese again rate as a number 1 choice. As an example: 6 to 8 young geese can control weed growth on 1 to 2 acres of strawberries.

If you are raising geese as a hobby and want a docile, pretty bird, then the medium-sized Pilgrim is a good choice. African and Toulouse geese with their colored feathers also are lovely "ornamental" geese and are good to eat.

If you want to exhibit your geese at poultry shows, choose a more exotic breed, such as Egyptian or Sebastopol; the competition will be less fierce than if you show Chinese or Toulouse geese.

If your interest in geese is more general and you want to have a small flock around—for pets, local color, an occasional dinner or some eggs, any of the breeds will do just fine.

But remember, no matter why you raise geese, or what breed you raise, *do not make a pet* of a bird that you will eventually eat! It would be very upsetting and probably impossible to slaughter, dress, and roast a family pet.

Chapter 12

HOUSING
REQUIREMENTS
FOR GEESE

Geese belong outdoors for the most part. But day-old goslings do need a dry, snug, draftfree shelter. They require this protective environment for at least the first month of their lives. Even full grown geese need a place to escape from the wind. You will notice that even wild geese hug the leeward bank of a lake or pond on raw, very windy days.

You can use a barn, chicken house, shed, garage, or old milk house for a goose house. The floor can be of wood, earth, or cement. A cement floor is the easiest to keep in sanitary condition, but also the coldest. However, you will provide bedding in the form of a litter made up of sawdust, wood shavings, chopped straw, crushed corn-cobs, or even dry peat moss to cover the floor.

Day-old goslings need a floor space of only about ¾ square feet each. When they are 2 weeks old they should have about 1½ square feet of space each. As they grow in size additional space should be provided. When geese are 4 to 6 weeks old, they are ready to go on range. By this time goslings need about 3 square feet of space each, when confined to the house during inclement weather.

It makes the most sense to provide for the geese's ultimate space requirements from the beginning. As a rule of thumb, 4 goslings need a minimum of 12 square feet, 8 birds need 24 square feet, and so on. You also need some room for a feed trough, waterer, and your own 2 feet, when walking about the pen and caring for them.

If you are going to use a large building to house the birds, you can partition off a section of it with plywood or chipboard. It does not have to be insulated but should be free of drafts. The southern-most exposure of the building is the best side to partition off, particularly if you live in a cold climate.

All windows should have a fine wire mesh fastened on the inside to keep wild birds from flying in and possibly bringing disease. Any holes or cracks in the walls or wooden flooring should be patched up or boarded over to prevent rats or weasels from getting in. These predators can slip through chicken wire of 1 inch in diameter, and they consider young goslings fair game. All doors should close tightly to prevent the family cat or dog from gaining entrance.

The flock of Buff geese are protected from predators by an electrified woven wire fence. (Photo courtesy USDA)

This range shelter is equipped with nest boxes.
(Photo courtesy USDA)

FENCING

By the time the geese are a month old, they are ready to forage for their feed in a pasture, yard, or garden. They still will need to be confined at night to protect them from predators and should be kept indoors if the weather is unusually rainy or cold.

One acre of good pasture can support 20 to 40 geese. Domestic geese are poor fliers, so a woven wire fence 3 feet high is sufficient to confine them to pasture. To protect the geese within the pasture from predators, a 5-foot-high fence should be constructed. Run a string of barbed wire a few inches above the ground outside the fence and another string of barbed wire on top of the woven wire. Although it is possible for the young goslings to escape through woven wire fencing, they will usually stay put if they have access to enough food, water, and shade.

If you don't have trees for shade, provide temporary shelters with scrap lumber and plywood. The sides of the shelter can be open, just provide a roof of some sort to enable the geese to get out of the sun.

NECESSARY EQUIPMENT

The equipment needed for raising geese is quite similar to that needed for chickens and ducks. You must provide some sort of brooder setup for the young goslings, with waterers, feed troughs, and a heat lamp. Older geese will require more waterers and feeders.

A 1-gallon fountain will provide plenty of water for a flock of 12 young birds. They should be able to immerse their whole bills in the waterer. It should be constructed (or have a guard) so that they cannot climb into it and become chilled or drown. Even better is to buy a 2-gallon water fountain which can serve them up to when they are fully grown.

Geese use a lot of water, especially when eating dry feed. You cannot keep the waterer clean at all times, but the supply should be plentiful. When the birds are grown, you can use cake tins, buckets, pails or old turkey roasting pans to water them.

A chicken trough will do just fine for feeding the goslings. In the beginning, a day-old gosling needs about ½ inch of feeding space at the trough. One 18-inch feed trough will provide plenty of room for at least a dozen birds from the time they are a day old until they are 4 to 6 weeks old and ready to go out on pasture. The trough should have a roller bar on top to prevent the birds from climbing inside and soiling the feed. Later on, when they are out on range, you can use larger covered troughs (to prevent rain from soaking the feed), or else distribute their feed on the ground in various places, so as to allow all of the birds to get something to eat.

If you are going to use a heat lamp to brood the goslings, you will need a 250-watt bulb complete with porcelain socket, reflector, and metal guard. You also will need a sturdy thermometer and a guard or corral, which will confine the birds to the source of heat, feed, and water for the first 10 days of their lives. The guard can be made of cardboard and should be at least 1 foot high. (The heat lamp setup and the guard are the same as for chickens. See pages 38–43 for more details.)

Chapter 13

CARING
FOR YOUR
FLOCK OF GEESE

Having selected the breed you want to raise, you are now ready to order your stock. You can start out by incubating eggs, buying day-old goslings, or buying mature birds. Each method has its pros and cons.

INCUBATING FERTILE EGGS

Fertile eggs ready for incubation from a reputable hatchery will cost about $1 each. There will be no guarantees to cover eggs that break in transit or fail to hatch. Novices can expect a 60 percent hatching rate. So if you want 8 geese, you will have to purchase 14 fertile eggs in order to come up with 8 live goslings. A no-frills artificial incubator will cost at least $35, and you will have to sprinkle the eggs each day, turn them, and check relative humidity and temperature for 29 to 31 days — the average period for hatching goose eggs.

Approximate cost: 14 fertile eggs $14
Manual incubator 35
$49

Projected net result: 8 live day-old goslings

Instead of buying a manual incubator, you can buy hens — either chicken, turkey, duck, or geese — to set the eggs (incubate) the natural way. Any chicken or turkey hens used for setting should be free of lice and mites. You will have to provide feed and water for the setting hens and isolate their nest in a quiet place for a period of 29 to 31 days. It may be difficult to obtain broody hens on short notice, and there is no guarantee that the hen will set the eggs through the full term.

Approximate cost: $14 fertile eggs.............. $14
2 chicken hens.......... 5
Feed for hens............... 3
Total $22

Projected net result: 8 live day-old goslings

BUYING DAY-OLD GOSLINGS

You can buy 8 day-old goslings of most breeds, guaranteed live arrival, for $40 or less from a reputable hatchery. And you don't have to scrounge around looking for a couple of licefree, broody hens to set the eggs for 29 to 31 days.

BUYING ADULT BIRDS

A popular mail order company specializing in poultry and pets offers the following selections in geese (the prices are for a pair of geese, a male and a female):

- **White Chinese** $65
- **Embden** 75
- **Pilgrim** 75
- **African** 85
- **Toulouse** 75
- **Sebastopol** 125
- **Egyptian** 85

Remember, that's just 2 geese for the money. At those prices, you certainly won't eat them. You hope they will mate, and the female will lay fertile eggs, and will set them until they hatch out, and then brood the goslings until they are on their own.

If there are farmers in your area who raise geese and will sell them, I'm sure that you can beat the prices quoted above. But, economically speaking, it will still cost more to buy adult birds than to go with day-old birds. And, you miss all the fun of watching them grow up.

I suggest that novices start with day-old goslings.

BUYING DAY-OLD GOSLINGS

If your main objective is meat and you want a breed that will achieve a good slaughter weight in the quickest time, I suggest you buy at least 8 Embden or White Chinese day-old as-hatched (unsexed) goslings. If fed heavily, the 8 birds will provide you with about 68 pounds of meat – giblets, neck, and bone included – within about 3½ months.

If you are interested in egg production, not necessarily for table fare, but perhaps to eventually raise your own goslings, the White Chinese is the best choice. I suggest you buy 8 to 12 birds, as-hatched.

Day-old geese are usually sold as-hatched. That is, half of the birds can be females and half males. When buying in small quantities, this does not always hold true. If you request either males or females from a hatchery, they will charge from $.50 to $1.00 extra per bird.

No matter what purpose you raise them for, I suggest that you begin with a minimum of 8 day-old goslings.

Where to Buy Your Goslings

Your local extension poultry specialist is an excellent source of help in choosing a hatchery or finding available birds. Country newspapers and farm magazines carry advertisements of waterfowl hatcheries. As a general rule, the closer to home you buy your birds, the better off you will be. Delivery charges will be less costly and it will be more convenient to contact your supplier for help and advice. Buy the best stock that you can afford, and this usually means dealing with a reputable hatchery.

When To Buy

Late spring or early summer is the best time for a beginner to buy day-old goslings and begin a goose raising venture. It depends, of course, on the climatic zone in which you live. In most of North America, June is a favorable time to have your young birds arrive. Temperate weather will aid in the success of your project. If possible, order your birds a month ahead of time.

PREPARING FOR THE ARRIVAL OF THE GOSLINGS

With the goslings on order, don't just sit back and wait for the postmaster to inform you of their arrival. Now's the time to order feed and prepare the goose house.

Preparing the Goose House

It is very important that the goslings' new home be as clean as possible. Sweep the ceiling and wash down the walls and floor with a disinfectant. If you are using an old chicken house, let it air out for 30 days after cleaning. Always follow the manufacturer's directions carefully when using disinfectants so that the goslings won't be harmed by any leftover chemical residues.

Day-old goslings are just about as helpless as day-old chicks. You will have to set up a brooder area for the goslings to provide them with plenty of warmth and ready access to feed and water.

For warmth, you can use a hover-type brooder, which is a canopy-shaped heating unit fired by gas, oil, or electricity. (See page 38 for an illustration.) A brooder for goslings should be positioned 3 or 4 inches higher above the floor than it would be positioned for chicks. Goslings are taller and larger than chicks.

A 250-watt heat lamp will provide all the warmth a small flock of birds requires and is less costly than a hover brooder. The lamp, complete with reflector and safety guard, should be hung in the cen-

These Toulouse goslings are about 1 month old and ready to be turned outside for a few hours every day. (Photo courtesy USDA)

ter of the pen, at a level of 18 to 24 inches above the floor. It can be suspended from rafters or the ceiling by wire, rope, or chain. The heat lamp arrangement must be adjustable so it can be raised or lowered during the course of brooding the goslings. Some growers use an infrared bulb in the lamp as it is thought that infrared light prevents cannibalism in the flock. (Cannibalism is the term used when 1 bird picks on another. When blood is drawn, other aggressive birds pick on the same unfortunate victim. The mutilated bird can be reduced to a helpless mess. Providing the flock with ample room, the correct feed, plenty of water, and proper heat all help to prevent cannibalism.)

Place the water and feed trough close to, but not directly under, the heat lamp. Then place the guard or corral around the lamp and troughs at a distance of about 4 feet away. The guard can be made of corrugated cardboard and should be about 1 foot high to contain the goslings and prevent drafts on the floor. (For an illustration of a chick guard, see page 43.) Be sure the guard is circular and has no corners. When goslings are startled they have a tendency to panic and all flee in 1 direction. If the guard has square corners, the goslings can pile up against the corner and injure themselves.

Next cover the floor with a 3-inch to 4-inch layer of litter. The litter keeps the floor warm and absorbs moisture. Any absorbent material will do: sawdust, peat moss, wood shavings, crushed corncobs, or finely chopped straw. Do not use materials with a slick surface, such as newspapers. The goslings can slip on smooth surfaces and injure their legs.

Finally, fasten or hang a thermometer next to the guard at about floor level. You will need to provide the goslings with a temperature of 85 to 90 degrees F.; the thermometer will help you monitor the goslings' comfort.

Set up your brooding arrangement about 1 day before the birds are due and make sure it works. Turn the lamp on, and, by lowering or raising it, find the right height so that you can maintain a constant temperature of 85 to 90 degrees F. within the pen.

Be sure to have feed on hand before the goslings arrive. Store it in a tight container so that rats and mice cannot get into it.

BROODING AND DAILY CARE OF THE GOSLINGS

It is important to be consistent in watering and feeding the birds, morning and night. When you do the chores, observe them closely. Are they all eating and drinking? Are they comfortable with the temperature?

At the start, you maintained a temperature in their house of 85 to 90 degrees F. By the second week, you can lower the temperature to 80 degrees F. by raising the heat lamp. Pick up any wet or soiled litter and replace it with fresh bedding. Keep the area around the waterer as dry as possible. Keep feed in front of them at all times and give them plenty of water. At this time they shouldn't need the guard anymore, and you can remove it from the pen.

By the third week you can lower the temperature in their house to 75 degrees F. They can be turned out into a yard on warm, sunny days to supplement their diet with weeds, seeds, grass, and bugs.

When they are 4 weeks old, the geese should be well-feathered out, and the temperature in their pen can be decreased to 70 degrees F. If you bought them in late spring or early summer as suggested, and the outdoor temperature during the daytime averages about 70 degrees F., you can turn the lamp off during the day. If nights are cool, turn the heat lamp on.

FEEDING

To start off, the goslings should be fed a ration of about 22 percent protein. A duck starter ration of small crumbles or pellets is best. If duck starter feed is not available, you can use a turkey or game bird ration.

If nothing else is available, you can use a baby chick starter of the correct protein level. Try to get a nonmedicated chick feed, as medicated chick feed can contain drugs not suitable for geese. Also, goslings grow much faster than chicks; their feed requirements are not alike. Improper nutrition can cause the goslings to develop leg problems. To help counteract any feed deficiencies, send the young birds out to a yard, pasture, or range, even if only for a couple of hours on nice sunny days. The goslings can start to spend time outdoors from age 2 weeks on. Put them back in their house if a storm threatens, and certainly do not leave them out overnight. Keep feed in their trough at all times, but only fill it about half full to prevent waste.

When the goslings are 4 weeks old, switch them to a duck grower ration of 15 to 18 percent protein. Again, if the duck feed is not available, you can substitute nonmedicated chick, turkey, or game bird rations of the correct protein level. If necessary, mash can be fed to goslings, but because of the shape of their bill and feeding habits, they waste a lot of it. And they need plenty of water at all times. If you have any starter ration left over, mix it in with the new feed and make the switch gradually. By this time, they should be actively grazing and eating less pre-packaged feed.

By the time they are 6 weeks old, the geese should be subsisting mainly on grass or weeds, with a light feeding of whole grains at night. Pen them inside at night for security. Do not give the geese access to dry feed within the house unless they also have plenty of water.

GEESE AS WEEDERS

Geese are excellent weeders for such cultivated crops as corn, cotton, sugar beets, and strawberries; and for nurseries, orchards, and vineyards. Young geese about 6 weeks old make the best weeders. The main idea is to keep them hungry. Turn them out in the morning and give them a light feeding of grain, corn, wheat, oats, or barley at night.

While in the field, they must have ample water and shade. Leafy trees, A-frames, or small shelters made of scrap lumber can provide the shade. Remove the geese from a strawberry patch when the fruit begins to ripen. Six to eight geese can do a good job of weeding an acre of strawberries.

GEESE ON PASTURE

Grass is a complete diet for geese during most of their life cycle. On a diet of grass, it will take them a little longer to reach the desired slaughter weight, but they won't be as fatty as birds that are semiconfined and full fed a high protein commercial feed their entire lives.

Geese prefer short succulent grasses and clovers. They will not eat tall dried-out mature grasses. One acre of good pasture can support 20 to 40 geese. Do not let them graze on a pasture that has been sprayed with poisonous chemicals, such as weed-killers.

Geese are handsome lawnmowers. The Goose Pond, a large public park in a town a couple of miles from where I live, harbors a population of about 300 wild Canada geese most of the year. No park service employees are needed to mow the grassy 5 acres that surround the pond. The geese keep the grass clipped short and luxuriant from spring until fall.

MANAGEMENT, FEED, AND GROWTH

How much commercial feed to provide for geese depends what you are aiming for. You may want to push your goslings as fast as possible in order for them to gain good slaughter weight. Or you may want to go at a slower pace and develop your birds for both meat and eggs.

"Green geese" is a term used to describe young geese that have been totally confined and full fed a ration high in protein and calories for their entire lives — a pretty short span that probably averages about 14 weeks. Goslings raised in this fashion by commerical growers can reach a live weight of 13 pounds in 14 weeks. They tend to be fattier than geese raised on range. In order to gain that much weight in such a short time, the confined goslings will each consume about 56 pounds of expensive feed.

In contrast, goslings that are reared on pasture or range, with a supplemental feeding of commercial rations or (cheaper) whole grains, can attain a live weight of about 12 pounds within the same 14 weeks, and yet will only consume about 38 pounds of feed each. In other words, the range-raised geese will weigh only about 1 pound less, but will eat about 18 pounds less of feed within the same 14 weeks. This can provide a decided saving in feed costs. Also, the birds raised outside should be healthier and have leaner carcasses.

On most farms with small flocks, geese are grown at a leisurely pace, and those destined for slaughter are marketed in late fall or winter, as they bring the best prices at Thanksgiving and Christmas. They will go through a molting stage earlier in the season, so they will be fairly free of troublesome pinfeathers. They will be 5 or 6 months old at this time and weigh 14 to 18 pounds, all depending on the breed.

Geese grow very rapidly until they are about 10 weeks old. Then they slow down. This, of course, affects feed efficiency.

POTENTIAL PROBLEMS

Predators are by far a bigger problem than disease when it comes to raising geese.

If you provide a secure house or shelter for the goslings for the first 4 weeks of their lives, there should be no predator problem. Family pets are among the worst offenders. Keep your dogs and cats away from the goslings. By the time they are a month old, geese are too big for most hawks to handle. If they are penned in at night, neither owls, weasels or foxes can get at them.

If your geese lay their eggs out-of-doors, crows, ravens, sea gulls, skunks, raccoons, and some snakes will eat them if given the chance.

I cannot think of any particular disease that is likely to affect a small farm flock. Geese are the most hardy of all domestic poultry.

Good sanitation is the best prevention for disease. When the geese are young and confined to the shelter, you should remove soiled and wet litter and replace it with dry fresh bedding. And always supply plenty of water.

If 1 or more of your birds becomes ill, consult your local extension poultry specialist. How can you tell if a bird is ill? It doesn't eat or drink and may suffer from a lack of balance or a twisted neck. Or it may have a respiratory infection, gasp and cough, and have no appetite. Always isolate a sick bird from the rest of the flock.

Geese do not usually have external parasites, such as lice and mites, and their lavish intake of water helps flush out any internal worms. It is important that their yard is well-drained and does not contain stagnant pools of water.

MANURE

The combination of manure and litter gathered from the floor of the goose house should be added to your compost pile for future use in your garden. Geese droppings are high in nitrogen, and fresh manure is not recommended for lawns or gardens.

PROCESSING MEAT BIRDS

When the geese are about 14 weeks old, it's time to consider butchering. Select a couple of the largest birds to weigh. If all the geese are to be butchered, it doesn't matter whether you catch males or females. However, if you are going to keep the females for breeding and egg laying, it is important to determine their sex.

If you are raising Pilgrim geese, the sexing is easy. The females will be a light gray and the males are all white. In breeds such as the Chinese and African, the males will have larger knobs on their heads. In regard to other less distinctive breeds, the adult males are usually larger in body, stand taller, and have longer necks. But when the geese are just 14 weeks old, it can be very difficult to distinguish between males and females. The only sure way is to examine their sex organs. First you have to catch the bird.

To Catch A Goose

Drive the bird into a corner of the pen or yard. Then grasp the neck firmly with one hand and with the other hand and arm, encircle its flapping wings and body and hug it against your waist. You can tuck its head under your elbow and support the bird with your hand and forearm underneath its breast and abdomen. Do not attempt to catch a goose by the legs as they are easily injured. If you don't entrap their wings, they can inflict some pretty smart blows to your head and face.

Determining Sex

The best way to sex a goose is to have an experienced person go through the genital examination of a goose and show you how it's done. In practice, the bird is held on its back under a bright light, with its head towards you and its tail pointing away. The tail is bent down and a finger is inserted through the vent and into the cloaca, the chamber which contains the reproductive organs. The finger is rotated gently in a circular fashion to relax the sphincter muscle. The opening becomes enlarged and the genital organs are exposed. You will find either the reproductive organ of the male or the eminence of the female.

Weigh the Birds

After knowing for sure that you have a male in hand, weigh the gosling. You can use a hanging scale or your bathroom scale. If the bird weighs less than 12 pounds, I suggest that you continue feeding him for another week or two. If he weighs at least 12 pounds, it's time to butcher him.

Slaughtering, Scalding, Eviscerating and Preparing the Bird

The goose to be butchered should be isolated in a crate or cage the night before it is to be killed. Give it plenty of water but no feed.

Killing the bird, scalding it, plucking the feathers, and dressing it is the same as processing a duck (see pages 123–126). But, you are now dealing with a bird about twice the size of a duck. The gosling has more feathers, much more down, and takes longer to pluck. Also the water temperature for scalding should be about 150 degrees F., and the bird should be immersed for 2 to 3 minutes. Sloshing the goose up and down in the scalding tub helps get the water through the feathers to the skin and aids in picking the bird. Again, as with ducks, wax can be used to remove the down. The bird should be dressed, chilled, and packaged just as you would process a duck.

Geese dress out to an average of 70 percent. Thus, a 12 pound bird will dress out to an average of 8½ pounds, with bones, neck, and giblets included.

Goose Feathers

It takes about 3 geese to provide 1 pound of feathers. The small feathers and down can be used to make vests, jackets, and bedding. Wash the feathers in lukewarm water with a bit of detergent added. Rinse them, wring gently, and spread them out to dry.

Goose down and feathers have commerical value. Companies that buy feathers are located in most large cities. Your extension poultry specialist can help you find potential buyers.

Chapter 14

BREEDING GEESE AND EGG PRODUCTION

If you want to build up your breeding flock, start selecting your prospective breeders at about 6 to 7 weeks of age. If you are raising only 8 birds, and 4 of them are females, you will probably keep all of them. If you have a choice, select the largest and most vigorous females. In addition, look for a good rate of growth. Later, as your flock grows and you gain experience with the individual geese, select also for good egg production, fertility, and hatchability, as well as size and vigor.

Although you don't need ganders for egg production, if you are looking for fertile egg production, you must have a few ganders around. Generally, 1 gander is sufficient for 3 to 5 females. Extra ganders can be raised to a good live weight of 12 to 14 pounds and slaughtered for meat.

Obviously, it is important to know how to distinguish the geese from the ganders. See page 167 on catching geese and determining the sex.

THE EGG PRODUCTION CYCLE

The goslings that you purchased in late spring or early summer will start laying eggs in the following year and continue to lay for several months. Generally, the geese will start laying eggs every other day in March and continue until June or July. They reach their peak production about 60 days after they commence to lay. Some light breeds, such as the White Chinese, will commence laying in late February.

If you live in the North, and there is no pasture or range in late fall and winter, you will have to feed your potential layers a light ration of grain to carry them through the winter. The geese, meanwhile, will prefer to stay outside, but they should have access to shelter during severe storms.

About a month before you expect egg production to begin, start feeding the geese a breeder or layer ration in pelletized form. These rations contain less energy than grower rations to prevent the birds from putting on fat. Fat birds produce fewer, smaller eggs. Most geese will lay 35 to 40 eggs per year.

Unlike chickens, geese lay more eggs during their second and third years than during their first year. They can lay for at least 10 years.

NEST BOXES

Although the laying geese don't have any special housing needs, you should provide nests to encourage them to lay their eggs where you can find them.

Build nest boxes inside the goose shelter or house. The nest should be about 24 inches square and 18 inches high. It does not require a top or a front. Fill the nest box with 3 to 4 inches of litter. One nest will serve for 3 females. If you don't provide nests for them, geese will sometimes make their own nests, forming beds of hay, grass, or straw, and lining them with feathers plucked from their own bodies.

You do not need to provide a pond or a body of water in order to have the geese mate and produce fertile eggs. Although they will go through a lot of awkward gyrations, geese can mate quite well on dry land.

EGG CARE

Eggs should be gathered at least a couple of times daily to encourage the geese to lay more. Imperfect, cracked, or soiled eggs should be used immediately for cooking and baking. Save only perfect eggs for future incubation. Eggs for incubation can be kept for up to 10 days at 55 degrees F. and a relative humidity of 75 percent before incubating.

INCUBATION

The incubation period for most geese is 29 to 31 days. Canada and Egyptian geese require 35 days. You can incubate goose eggs naturally. Embdens and Pilgrims are more likely to be good mothers than most other breeds. A goose will cover 9 or 10 eggs at a time. Don't disturb the nest, and see that plenty of feed and water are available nearby for the setting bird.

Part Four: Appendixes

Appendix A

GLOSSARY

Broiler-fryer. A chicken less than 3 months old, male or female, with pliable, smooth-textured skin and tender meat. The bird's cartilage is quite flexible.

Capon. A castrated male chicken, used for roasting. The neutering process makes for a bird with soft, smooth skin and tender meat. Capons usually are butchered when they are 8 months old or less and weigh 5 pounds or more.

Cockerel. A male chicken less than 1 year old.

Cull. To separate or remove from the flock birds of inferior quality. Also, a bird thus removed.

Duck. Generally speaking, the family of birds that swim; in particular, the female duck as distinguished from the male (drake).

Duckling. A young or immature duck, either male or female.

Drake. A mature male duck.

Gander. A mature male goose.

Geese. A general term covered males and females (goose and gander) and all ages.

Gizzard. The second stomach of birds. This organ has thick walls and a horny, muscular lining to aid in the grinding down of food, particularly whole grains.

Goose. A subfamily of waterfowl; swimming birds. Intermediate between ducks and swans. The female, as distinguished from the male (gander).

Gosling. An immature goose of either sex.

Grit. Tiny particles of stone that do not dissolve in liquid.

Hackle Feathers. Neck feathers often used in making fishing lures.

Hen. A mature female chicken, at least 1 year old. Also, a mature female duck.

Molt. The natural process of shedding or casting off old feathers and growing new ones.

Poultry. A general term designating domesticated birds such as chickens, ducks, geese, turkeys, guineas, pigeons, swans, and peafowl.

Pullet. An immature female chicken less than 1 year old.

Roaster. A chicken of either sex, customarily 5 months old or less, still tender-meated, having fairly flexible breastbone cartilage, and weighing 5 pounds or more. (By the time a chicken is a year old, the tip of the breastbone becomes hard and inflexible.)

Rooster. Also called a cock. A mature male chicken at least 1 year old. Not recommended for eating as the skin is coarse and the meat is tough.

Waterfowl. A term applied to birds that swim, including ducks and geese.

Appendix B

PLAN FOR A PORTABLE CHICKEN COOP

Edward N. Robinson, author of the classic homesteader's manual, *The Have-More Plan*, has designed an ingenious movable chicken coop, just big enough for 8 laying hens. This coop is inexpensive to build (approximately $280 in materials plus $150 for the trailer), attractive, and easily cleaned.

With 2 windows and a door, the hens get plenty of fresh air, and Ed has easy access inside the coop and nests just by reaching in. The coop can be oriented to face south to provide the hens with plenty of solar heat.

The floor of the coop is turkey wire, so the hens droppings are instantly removed. The coop can be moved around the yard and garden to provide labor-free fertilizing.

Sides of reinforced mesh drop down from the trailer to provide an instant yard for the chickens to scratch in to their hearts' delight. The chickens have access to the yard via a trap door which is lowered from the outside by pulling on the spring bolt that supports it.

The coop can be used year-round. In the winter, Ed puts corrugated cardboard over the wire floor and adds 3 to 4 inches of hay or straw on top to keep the hens warm. The cardboard and hay are replaced when they become soiled.

The coop is light enough to be easily pulled by hand around the yard. It can also be hitched to a car and moved elsewhere. Ed takes advantage of the portable coop by moving it to an agreeable neighbor's yard when he wants to get away for the weekend—a feasible solution to the problem of finding someone to do chores when you want to take a holiday.

MATERIALS LIST

Lumber	AMOUNT	SIZE
	6 boards	2" × 4" × 12' treated pine (ripped into 2×2s for framing)
	6 boards	2" × 4" × 16' treated pine (ripped into 2×2s for framing)
	4 pieces	¼" × 4' × 8' Lauan pine plywood (exterior and interior sheathing)
	2 boards	2-5/4" × 6" × 12' treated pine (ripped into 3" widths. Used to fasten both sides of front frame to extend the width of the floor to the outside of the wheels, to allow wire fencing to drop down.)
	32 board feet	1" × 4" common pine (ripped into 1" × 2" framing for roosts and rests on wire floor.)
	1 piece	1½ to ½" × 4' × 8' exterior grade plywood (roof)

Hardware	AMOUNT	SIZE
	2	3" × 3" brass hinges (for ramp)
	4	brass hook-and-eyes (for doors, windows, fencing)
	30	loose nuts, bolts, screws
	1	bolt catch (for ramp)
	3	5" × 10" galvanized metal eaves
		4d galvanized box nails
	8	3" × 3" weatherproof brass hinges (for drop-down fence)
		staples
		3" sheetrock screws
	4	3" brass barrel bolts (for egg door and hinged fencing members)

Insulation   3½" × 15' fiberglass (cut to fit wall cavity)

Roofing   108 square feet black roofing paper

Plexiglass   2 24" × 30" × 3/16" (for exterior of hinged windows and door)

Wire   1 6' × 4' piece of 1" × 2" turkey wire (for floor)
1 20' roll 2'-high poultry netting (1" size) (for drop-down fencing)

Trailer   1 trailer kit (frame, wheels, tow bar, lights etc.)

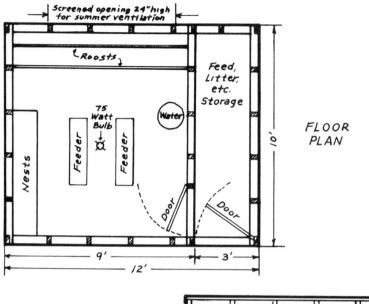

Screened opening 24"high
for summer ventilation

Roosts

Feed,
Litter,
etc.
Storage

75
Watt
Bulb

Water

Feeder

Feeder

Nests

Door

Door

FLOOR
PLAN

10'

9'

3'

12'

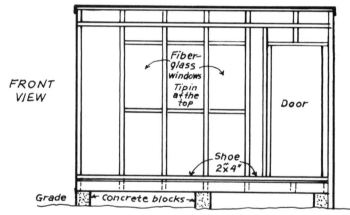

FRONT
VIEW

Fiber-
glass
windows
Tip in
at the
top

Door

Shoe
2"x 4"

Grade

Concrete blocks

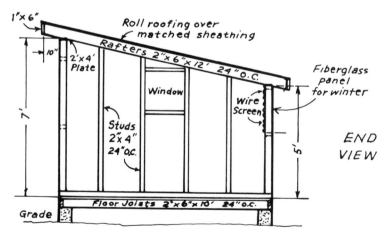

1"x 6"

Roll roofing over
matched sheathing

10"

Rafters 2"x 6"x 12' 24"O.C.

2'x 4'
Plate

Window

Fiberglass
panel
for winter

Wire
Screen

Studs
2"x 4"
24"O.C.

END
VIEW

7'

5'

Floor Joists 2"x 6"x 10' 24"o.c.

Grade

Appendix C

PLAN FOR A PERMANENT CHICKEN COOP

This chicken coop was designed by the Cooperative Extension Service, Agricultural Engineering Department, University of New Hampshire.

MATERIALS LIST

Lumber	AMOUNT	SIZE
	7 boards	2" × 6" × 10' (floor joists)
	2 boards	2" × 6" × 12' (front and rear sills)
	54 linear feet	2" × 4" (shoe)
	9 boards	2" × 4" × 5' (rear studs)
	9 boards	2" × 4" × 7' (front studs)
	4 boards	2" × 4" × 12' (end studs)
	2 boards	2" × 4" × 12' (partition studs)
	2 boards	2" × 4" × 12' (plates)
	4 boards	2" × 4" × 12' (miscellaneous framing)
	150 board feet	T&G sheathing (floor)
	450 board feet	T&G sheathing (siding and doors)

Miscellaneous   6 concrete blocks: 8' × 8' × 16'
2 pieces 2' × 10' flat fiberglass (windows)
1½ sq. roll roofing
nails and hardware

Appendix D

POULTRY PUBLICATION AND PERIODICALS

These periodicals provide a wealth of information on chickens, ducks, and geese.

Backyard Poultry
Route 1, Box 7
Waterloo, Wisconsin 53594
12 issues for $10
A down-to-earth monthly magazine aimed toward the small flock owner and chock full of good information and helpful tips in regard to raising all types of poultry. Lots of good ads.

Hen House Herald
PO Box 1011
Council Bluffs, Iowa 51502
$6.50 per year
A monthly newspaper promoting poultry and small animals.

Poultry Press
Box 947
York, Pennsylvania 17405
$8.00 per year
This publication has been promoting standard-bred poultry for 70 years. Big on poultry exhibitions and information for fanciers. Good source of birds for small flocks.

ADDITIONAL PUBLICATIONS AND REFERENCES
Animal Science
M. E. Ensminger. (The Interstate Publishing Company Danville, Ill) 1969. This book of over 1200 pages is considered the bible of animal husbandry. It devotes 175 pages to the poultry industry and contains a wealth of information about raising broiler-fryers and egg-laying chickens.

Duck and Goose Raising
Publication 532, H. L. Orr (Ontario Ministry of Agriculture and Food, Department of Animal and Poultry Science, Ontario Agriculture College, Guelph, Ontario, Canada) 1981.

Eggcyclopedia
American Egg Board. (1460 Renaissance Drive, Park Ridge, IL 60068) 1981. This 50-page booklet provides a well-done potpourri about eggs and their source, the chicken.

Farm Poultry Management
Farmers Bulletin #2197, USDA (Superintendent of Documents US Government Printing Office, Washington, DC 20402. 1977 A brief 37-page bulletin which contains good information but is short on details.

National Poultry Improvement Plan
USDA (Beltsville, MD 20705) 1984. The N.P.I.P. is a plan in which the USDA cooperates with state authorities in the administration of regulations for the improvement of poultry, poultry products, and hatcheries. The N.P.I.P. issues a series of booklets that tell you which participating hatcheries are US Pullorum-Typhoid (disease) clean. They also issue a directory of participating hatcheries which handle chickens, turkeys, waterfowl, exhibition poultry, and game birds.

Raising Ducks
Farmers Bulletin #2215, USDA. (Superintendent of Documents, US Government Printing Office, Washington, DC 20404) 1976.

Raising the Home Duck Flock
David Holderread. (Holderread Waterfowl Farm and Hatchery, PO Box 492, Corvallis, OR 97339) 1979.

Raising Geese
Farmers Bulletin #2251, USDA. (Superintendent of Documents, US Government Printing Office, Washington, DC 20404) 1978 (rev.).

The Vital 10% for Poultry
Grass Yearbook of Agriculture, H. R. Bird. (USDA, Beltsville, MD 20705) 1948.

Appendix E

COMMERCIAL HATCHERIES

The hatcheries listed below represents a random sampling of hatcheries across the United States. These hatcheries are not necessarily recommended, nor are these hatcheries the only sources of chicks, ducklings, and goslings.

Murray McMurray Hatchery
A123
Webster City, Iowa 50595
This hatchery sells only baby chicks and offers a beautiful free color catalog.

Stromberg's Chicks & Pets Unlimited
Pine River 5
Minnesota 56474
You can buy anything from hatching eggs to a buffalo from Stromber's. They also sell baby chicks and matched pairs and trios of adult birds. They offer a 56-page color catalog for $1.

The Chicken Lady
33530 Orange Street
Lake Elsinore, CA 92330
This hatchery offers baby chicks, ducklings, and turkey poults.

Heart of Missouri Poultry Farm
Box 954B
Columbia, MO 65205
Columbia, MO 65205
Sells baby chicks, ducklings, goslings, and turkey poults. Send $1 for an illustrated brochure, deductible from order.

Ideal Poultry Breeding Farms, Inc.
Box 591 C
Cameron, TX 76520
Offers many varieties of baby chicks. Write for price list.

Inman Hatcheries
Box 616
Aberdeen, SD 57401
Offers baby chicks, ducklings, goslings, and turkey poults. also sells hatching eggs and started chicks. Send $1 for catalog, deductible from order.

Hoffman Hatchery, Inc.
Gratz, PA 17030
Sells baby chicks, ducklings, goslings, and turkey poults. Free brochure on request.

Mt. Healthy Hatcheries
Mt. Healthy, OH 45231
Offers baby chicks, ducklings, goslings, and turkey poults. Free brochure on request.

Ridgway Hatcheries, Inc.
Box 306
LaRue, OH 43332
Sells baby chicks, ducklings, goslings, and turkey poults. Free brochure on request.

Appendix F

SOURCES OF EQUIPMENT AND SUPPLIES

This listing is not complete and does not imply any recommendation. It is provided as an aid.

The Chicken Lady
33530 Orange Street
Lake Elsinore, CA 92330

Countryside General Store
103 N. Monroe Street
Waterloo, WI 53594

GQF Manufacturing Company
PO Box 1552-BP
Savannah, GA 31498

Hoffman Hatchery, Inc.
Gratz, PA 17030

Marsh Manufacturing, Inc.
7171 Patterson Drive
Garden Grove, CA 92641

Shoemaker Poultry Supply
PO Box 331
Mt. Gilead, OH 43338

Stromberg's
Pine River 5, MN 56474

Appendix G

MISCELLANEOUS SOURCES

American Poultry Association, Inc.
Bertha Traver, Secretary
RD 4, Box 351
Troy, NY 12180
   This organization is devoted exclusively to the poultry industry and establishes breed standards. Members received a newsletter and yearbook. The APA sponsors poultry exhibitions.

CORRESPONDENCE COURSES
The Pennsylvania State University
307 Agricultural Administration Building
University Park, PA 16802
   Write for a catalog of courses offered.

Appendix H

ADDRESSES OF EXTENSION SERVICE OFFICES BY STATE

ALABAMA
Auburn University,
Auburn, Alabama 36830

102 Federal Bldg.
P.O. Box 15
Cullman, Alabama 35055

ARIZONA
Agricultural Science Building
University of Arizona
Tucson, Arizona 85721

ARKANSAS
University of Arkansas
P.O. Box 391
Little Rock, Arkansas 72203

University of Arkansas
Dept. Animal Science
R-C123
Fayetteville, Arkansas 72701

CALIFORNIA
University of California
Department of Avian Sciences
Davis, California 95616

University of California
306 Agricultural Extension Building
Riverside, California 92502

P.O. Box 1411
Modesto, California 95353

Agr. Res. & Ext. Center
9240 S. Riverbend Ave.
Parlier, California 93648

Suite 202, 21160 Box Springs Rd.
Riverside, California 92507

566 Lugo Avenue
San Bernardino, California 92415

Building 4, 5555 Overland Avenue
San Diego, California 92123

Room 100-P, Mendocino Avenue
Santa Rosa, California 95401

684 Bueno Vista St.
Ventura, California 93001

1000 S. Harbor Blvd.
Anaheim, California 92805

COLORADO
Colorado State University
Fort Collins, Colorado 80521

CONNECTICUT
University of Connecticut
Storrs, Connecticut 06268

Agricultural Center
Haddam, Connecticut 06492

24 Hyde Avenue Rt. 30
Rockville, Connecticut 06066

562 New London Turnpike
Norwich, Connecticut 06360

Appendix H (continued)

DELAWARE

University of Delaware
RD 2, Box 48
Georgetown, Delaware 19947

Agr. Hall
Newark, Delaware 19711

FLORIDA

University of Florida
Gainesville, Florida 32603

Chipley, Florida 32428

GEORGIA

University of Georgia
Athens, Georgia 30602

Calhoun, Georgia 30701

Oakwood, Georgia 30566
Tifton, Georgia 31794

HAWAII

University of Hawaii
1825 Edmondson Rd.
Honolulu, Hawaii 96822

IDAHO

University of Idaho
Moscow, Idaho 83843

ILLINOIS

University of Illinois
322 Mumford Hall
Urbana, Illinois 61801

INDIANA

Purdue University
Lafayette, Indiana 47907

IOWA

Iowa State University
Kildee Hall
Ames, Iowa 50010

KANSAS

Kansas State University
Leland Call Hall
Manhattan, Kansas 66506

KENTUCKY

University of Kentucky
Lexington, Kentucky 40506

Somerset Community College
808 Monticello Rd.
Somerset, Kentucky 42501

1270 Montgomery Avenue
Ashland, Kentucky 41101

LOUISIANA

Louisiana State University
Knapp Hall
Baton Rouge, Louisiana 70803

MAINE

University of Maine
Hitchner Hall
Orono, Maine 04473

P.O. Building
Belfast, Maine 04915

P.O. Building
Lewiston, Maine 04241

Federal Bldg. Rm. 209
Rockland, Maine 04841

MARYLAND

University of Maryland
Dept. of Poultry Science
College Park, Maryland 20742

Broiler Sub-Station,
RFD 5
Salisbury, Maryland 21801

MASSACHUSETTS

University of Massachusetts
314 Stockbridge Hall
Amherst, Massachusetts 01002

MINNESOTA

University of Minnesota
St. Paul, Minnesota 55101

MICHIGAN
Michigan State University
113 Anthony Hall
East Lansing, Michigan 48823

Coop Extension Service
P.O. Box 79
Zeeland, Michigan 49464

MISSISSIPPI
Mississippi State University
P.O. Box 5425
Mississippi State, Mississippi 39762

Box 9714
Jackson, Mississippi 39206

MISSOURI
University of Missouri
Columbia, Missouri 65201

RD1 Bldg.
Belcrest & E. Trafficway
Springfield, Missouri 65802

NEBRASKA
University of Nebraska
Lincoln, Nebraska 65803

NEW HAMPSHIRE
University of New Hampshire
55 Pleasant St., Rm. 331
Concord, New Hampshire 03301

Kendall Hall
Durham, New Hampshire 03824

NEW JERSEY
Rutgers – The State University
CAES P.O. Box 231
New Brunswick, New Jersey 08903

NEW MEXICO
New Mexico State University
Dept. Poultry Sci., Box 3P
Las Cruces, New Mexico 88001

NEW YORK
Cornell University
Rice Hall, Ithaca, New York 14850
249 Highland Ave.
  Rochester, New York 14620
380 Federal Building
  Syracuse, New York 13202
Coop Ext. Regional Office, Martin Rd.
  Voorheesville, N.Y. 12186

NORTH CAROLINA
North Carolina State University
P.O. Box 5307 Scott Hall
Raleigh, N.C. 27607

NORTH DAKOTA
North Dakota State University
Stevens Hall, Rm. 218A
Fargo, North Dakota 58102

OHIO
Ohio State University
2120 Fyffe Road
Columbus, Ohio 43210

OKLAHOMA
Oklahoma State University
Stillwater, Oklahoma 74074

OREGON
Oregon State University
Poultry Sci. Dept.
Corvallis, Oregon 97331

PENNSYLVANIA
Pennsylvania State University
University Park, Pennsylvania 16802

PUERTO RICO
University of Puerto Rico
Rio Piedras, Puerto Rico 00928

RHODE ISLAND
University of Rhode Island
Kingston, Rhode Island 02881

SOUTH CAROLINA
Clemson University
Clemson, South Carolina 29631
P.O. Box 378,
  York, South Carolina 29745
P.O. Box 1711
  Columbia, South Carolina 29201

SOUTH DAKOTA
South Dakota State University
Brookings, South Dakota 57006

Appendix H (continued)

TENNESSEE
University of Tennessee
P.O. Box 1071
Knoxville, Tennessee 37901

TEXAS
Texas A & M University
College Station, Texas 77843

Res. & Ext. Center
Box 220
Overton, Texas 75684

UTAH
Utah State University
Logan, Utah 84322

VERMONT
University of Vermont
Burlington, Vermont 05401

VIRGINIA
Virginia Polytechnic Institute
Blacksburg, Virginia 24061

WASHINGTON
Washington State University
Pullman, Washington 99163

Western Washington, Res. & Ext. Center
Puyallup, Washington 98371

WEST VIRGINIA
West Virginia University
Morgantown, West Virginia 26506

WISCONSIN
University of Wisconsin
Dept. Poultry Sci., Col. of Agr. & Life Sci.
1675 Observatory Dr.
Madison, Wisconsin 53706

WYOMING
University of Wyoming
Univ. Station, P.O. Box 3354
Laramie, Wyoming 82071

WASHINGTON, D.C.
Extension Service – USDA
Room 5509 South Building
Washington, D.C. 20250

186 • RAISING POULTRY SUCCESSFULLY

Index

Numbers in italics refer to illustrations.

Index (continued)

Index (continued)

Meat
chicken, 14–17, *16*, 20, 23–62
comparison of cooked beef and, 11
duck, 102–3
geese, 145
Mites, 87
Molting
chicken, 82–83
forced, 83
ducks, 122
Muscovy, 110–11

N
National Poultry Improvement Plan, 35
Nest boxes, 133, *133*, 170
Nests, 74–75
New Hampshires, 19, 20, 66
Newcastle disease, 88–89
Northern fowl mites, 87

P
Parasites, 86–87
Partridge Rocks, 20
Pasture
ducks, 137
geese, 147, 165
Peepers, *52*, 53
Pilgrim, 152, *152*, 152
Pinfeathers, 122
Plucking feathers, 58, 123–24
Ponds, 131
Predators, 36–37, 141, 166
Production Reds, 20
Pullets, buying, 65–69

R
Raising
chickens, reasons for, 9–12. *See also* Chicken(s)
different poultry types together, 141
ducks, reasons for, 101–6. *See also* Duck(s)
geese, reasons for, 145–48. *See also* Geese
Red Leghorns, 20
Red mites, 87
Rhode Island Reds, 19, 20, 66
Roasters, 20. *See also* Chicken(s); Meat, chickens
Rock-Cornish. *See* Cornish-Rock
Roosts, 72–73
Rouen, 110
Roundworm, 87

S
Scalding, 58, 123–24, 168
Sebatopol, 153
Sex, determining, 138–39, 167
Shaver, 33
Silver Laced Wyandotte, 19, 66
Silver Leghorns, 20
Skinning, 124
Specs, *52*, 53
Starve-outs, 35
Straight-run chicks, 34
Streams, 131

T
Toulouse, 153
goslings, *162*
Turken, 17

U
US Pullorum-Typhoid clean, 35
US standards for quality of individual shell eggs, 96
US weight classes for consumer grades for shell eggs, 94
USDA, 35

V
Vaccination, 88–89
Vantress, 33

W
Waterers, 41, *41*, 71–72, 132–33
Weeders, geese as, 165
White Chinese, *150*, 151
White Cochins, 20, 21
White Cornish, 66
White Leghorns, 18, *18*, 20, 66
White Pekin, 109, *108*
White Plymouth-Rock, 19, 66
White Rocks, 20
Whole grains, 56, 79
Wyandottes, 18, 20

Books from

WILLIAMSON PUBLISHING

The Sheep Raiser's Manual
by William Kruesi
"Overall, *The Sheep Raiser's Manual* does a better job of integrating all aspects of sheep farming into a successful sheep enterprise than any other book published in the United States."

Dr. Paul Saenger
New England Farmer

288 pages, 6 × 9, illustrations, photos, charts & graphs.
Quality paperback, $13.95.

Raising Pigs Successfully
by Kathy and Bob Kellogg
Everything you need to know for the perfect low-cost, low-work pig raising operation from choosing piglets, to housing, feeds, care, breeding, slaughtering, packaging, and even cooking your home grown pork.

224 pages, 6 × 9, illustrations and photos.
Quality paperback, $9.95.

Raising Rabbits Successfully
by Bob Bennett
Written by one of the foremost rabbit authorities, this book is ideal for the beginner raising rabbits for food, fun, shows and profit.

192 pages, 6 × 9, illustrations and photos.
Quality paperback, $9.95.

Home Tanning & Leathercraft Simplified
by Kathy Kellogg
"An exceptionally thorough and readable do-it-yourself book."
Library Journal

192 pages, 6 × 9, step-by-step illustrations, photos, tanning recipes.
Quality paperback, $9.95.

Books from

WWILLIAMSON PUBLISHING

Practical Pole Building Construction
With Plans for Barns, Cabins & Outbuildings
by Leigh Seddon
　　Complete how-to-build information with original architectural plans for small barn, shed, animal shelter, horse stall, as well as cabins and home.

176 pages, 8½ × 11, over 100 architectural renderings, charts, photos.
Quality paperback, $10.95.

At your bookstore or
To order directly from Williamson Publishing, send check or money order to **William Publishing Co., Church Hill Road,**
　　　　P.O. Box 185, Charlotte, Vermont 05445.
Please add $2.50 for postage and handling. For toll-free ordering, call 800-234-8791.